Environmental Footprints and Eco-design of Products and Processes

Series Editor

Subramanian Senthilkannan Muthu, Head of Sustainability - SgT Group and API, Hong Kong, Kowloon, Hong Kong

Indexed by Scopus

This series aims to broadly cover all the aspects related to environmental assessment of products, development of environmental and ecological indicators and eco-design of various products and processes. Below are the areas fall under the aims and scope of this series, but not limited to: Environmental Life Cycle Assessment; Social Life Cycle Assessment; Organizational and Product Carbon Footprints; Ecological, Energy and Water Footprints; Life cycle costing; Environmental and sustainable indicators; Environmental impact assessment methods and tools; Eco-design (sustainable design) aspects and tools; Biodegradation studies; Recycling; Solid waste management; Environmental and social audits; Green Purchasing and tools; Product environmental footprints; Environmental management standards and regulations; Eco-labels; Green Claims and green washing; Assessment of sustainability aspects.

More information about this series at http://www.springer.com/series/13340

Subramanian Senthilkannan Muthu
Editor

LCA Based Carbon Footprint Assessment

 Springer

Editor
Subramanian Senthilkannan Muthu
Head of Sustainability - SgT Group and API
Kowloon, Hong Kong

ISSN 2345-7651   ISSN 2345-766X   (electronic)
Environmental Footprints and Eco-design of Products and Processes
ISBN 978-981-33-4375-7   ISBN 978-981-33-4373-3   (eBook)
https://doi.org/10.1007/978-981-33-4373-3

This Springer imprint is published by the registered company Springer Nature Singapore Pte Ltd.
The registered company address is: 152 Beach Road, #21-01/04 Gateway East, Singapore 189721,
Singapore

This book is dedicated to:

The lotus feet of my beloved Lord Pazhaniandavar

My beloved late Father

My beloved Mother

My beloved Wife Karpagam and Daughters

—Anu and Karthika

My beloved Brother

—Raghavan

Everyone working in various industrial sectors to reduce the carbon footprint to make our planet earth SUSTAINABLE

Contents

About the Editor

Dr. Subramanian Senthilkannan Muthu currently works for SgT Group as Head of Sustainability, and is based out of Hong Kong. He earned his PhD from The Hong Kong Polytechnic University, and is a renowned expert in the areas of Environmental Sustainability in Textiles & Clothing Supply Chain, Product Life Cycle Assessment (LCA) and Product Carbon Footprint Assessment (PCF) in various industrial sectors. He has five years of industrial experience in textile manufacturing, research and development and textile testing and seven years of experience in Life Cycle Assessment (LCA), carbon and ecological footprints assessment of various consumer products. He has published more than 75 research publications, written numerous book chapters and authored/edited over 85 books in the areas of Carbon Footprint, Recycling, Environmental Assessment and Environmental Sustainability.

Carbon Footprint Assessment with LCA Methodology

Gaurav Jha, Shatrughan Soren, and Kapil Deo Mehta

Abstract Extended carbon footprints are causing damage to nature and their systems propagating a series of disastrous events. These extending footprints are due to the nonsustainable industrial practices, use of fossil fuels, improper disposal, and waste management, etc. Therefore, an assessment becomes a must prior to the mitigation efforts. The chapter deals with the carbon footprint and its assessment using LCA. During which, several other topics such as carbon footprint, environmental concerns, and their mitigation, need for assessment, LCA for various methods, and LCA tools, are discussed in detail. The chapter outlines the efficacy of LCA for footprint assessment justified by the study and investigations carried out worldwide, delivering attainability to a detailed description of LCA in the assessment.

Keywords Carbon-footprint · LCA · Environmental aspects · Sustainability · Metallurgical processes · LCA tools · Climate change · Greenhouse gases

1 Introduction

Carbon footprint is the carbon emissions being released from human activities [1–3]. These activities are boundless, either say it industries, lavish lifestyle, use of fossil fuels, not following the regulations, etc. [4, 5]. One of the major contributors is the industries that are operating with carbon-based fuels. These fuels are the direct source of carbon emissions. The major industries which depend on the coal/coke are the metallurgical and the thermal power sectors, out of which iron and steel sectors are the major emissions generating operations that are contributing approximately 60% of the total carbon emissions [6]. In the present time, these sectors depend

G. Jha (✉) · S. Soren
Department of Fuel Minerals and Metallurgical Engineering, Indian Institute of Technology (Indian School of Mines), Dhanbad 826004, India
e-mail: gjha29311@gmail.com

K. D. Mehta
Mineral Processing Division, National Metallurgical Laboratory, Jamshedpur 831007, India

© The Author(s), under exclusive license to Springer Nature Singapore Pte Ltd. 2021
S. S. Muthu (ed.), *LCA Based Carbon Footprint Assessment*, Environmental Footprints and Eco-design of Products and Processes, https://doi.org/10.1007/978-981-33-4373-3_1

on conventional fossil fuels like coal and coke. Although the scientific fraternity is recommending the usage of renewable biomass fuels in place of fossil fuels, that requires a due course and extensive research. Before implicating such a big change in the present system, it is essential to identify the present impact of coal/coke usage on the environment [7–10]. This is where carbon emission assessment necessitates for metallurgical applications. The metallurgical sector is not the only cause; various other industries also contribute to the global footprints. The chapter will assess various processes that are major carbon footprint contributors with the help of a life cycle methodology. The questions such as; how to perform an assessment, where it executes, and with what obvious tools, will get answered, as the chapter will proceed further. The information presented in the chapter will be felt more intriguing when the readers will find the interconnection among environmental and technological goals referring to the carbon footprint. The chapter can be read with the presented roadmap highlighting a detailed interest of the individual parts. The following are the details of the individual sections.

Chapter 2—Carbon footprint: The section discusses the carbon footprint, their origin, critical concerns, and current trends.

Chapter 3—Environmental concern and regulations for carbon footprint: Carbon footprint comes up with a variety of long-term environmental concerns, which are being highlighted in the section. The section is tried to maintain short without compromising with the key information.

Chapter 4—Probable mitigation of these concerns: The concerns raised by increased footprints are getting severe with time and that has to be stopped with sustainable developments. The section presents certain points that may help in reducing these footprints.

Chapter 5—Need for an assessment: Carbon footprint needs an assessment prior to their mitigation stages because it may face difficulties when you do not have proper limits. There comes the need for the assessment. The section highlights the background and interest requiring assessment.

Chapter 6—Lifecycle assessment: The section discusses LCA and their methodology details, providing following information (a) evolution of LCA and the purpose of the study, (b) code for impactful conduct of LCA, (c) learning objectives of LCA, and (d) types of LCA.

Chapter 7—LCA methodology in terms of carbon footprint assessment: The section presents an LCA assessment for the global carbon footprint including all the details of goal and scope, inventory generation, impact assessment, and interpretation.

Chapter 8—Other LCA examples: The section integrates various other assessments for industries like construction, textiles, metallurgical operations, municipal wastes, etc.

Chapter 9—Tools and software's used for LCA: The section discusses the various software solution to pursue assessment.

Chapter 10—**Summary**: At last comes the summary, which comprises the outcomes of the chapter highlighting the efficacy of LCA for carbon footprint assessment purposes.

2 Carbon Footprint

The term carbon footprint has a very clear meaning but a rather arduous understanding. The term has established first from an ecological footprint, which was developed by William E. Rees and Mathis Wackernagel in the 1990s [11]. It was calculated as tonnes per year CO_2 equivalent. It basically represents the footprint of carbon emissions derived from human activity. A carbon footprint is a prominent component of the ecological footprint and must be an addressable concern. This carbon footprint is usually a common term because it is imposed by almost every component with carbon or which runs on carbon-based components. It is not only that the sole contributor to carbon footprint is, industries, appliances, vehicles, and fuel applications but even our activities are contributing that is termed as per capita carbon footprint. As per the statistics, on average, Australia has a per capita footprint of 17 tonnes followed by the USA with 16.2 tonnes and Canada with 15 tonnes whereas this average is 4.8 tonnes globally (for 2017) [12, 13]. In other words, somehow, we are also contributing to carbon footprint through our activities which we do not consider us accountable for. The discussion to make anyone accountable while sitting in an air-conditioned room is of no advantage. If you will look at the share of the world's carbon footprint, China is the top contributor with a 27.2% share in footprint and that is imposed by 19.9% of the world population. On the other hand, the USA which has 4.5% of the world population adds 15.6% of the total footprint and stands in second place. India is the second most populated country with 17.5% of the world population that contributes 6.3% share in footprint (see Fig. 1). The path is followed by Russia, Japan, and Germany [14]. One can observe the dependency of the country's development on the carbon footprint share. Developed countries have a higher carbon footprint than developing ones.

Also, when looking at the distribution among the various sectors, electricity generation and heat production sectors are the major contributors with a 25% share followed by the agriculture sector (cultivation and deforestation) with a 24% share. The third major sector is the industry with a 21% share (IPCC 2014) followed by transportation, building, and others with a respective share of 14%, 6%, and 10%. Out of all the sectors, the major contributor is fossil fuels adding up a huge footprint with 65% share. Deforestation and decay of biomass add up 11% whereas 16% from methane, 6% from NOx, and 2% from CFC/HFC/PFCs (fluorinated gases). The 65% share by fossil fuels was extensively derived from the metallurgical and power generation sector. The annual generation of carbon emissions from fossil fuels is huge and imposes a critical threat. When looking at the spatial distribution of carbon footprint, major oil-producing countries are the ones having the world's largest per capita CO_2 emitters (from fossil fuels). Despite having a very low population, these

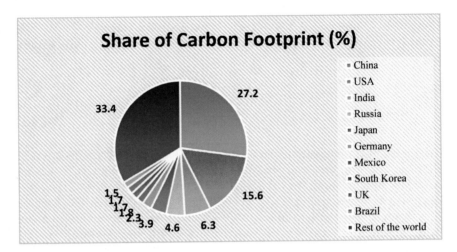

Fig. 1 Worldwide share of carbon footprint [14]

countries are producing a huge emission. Most of them are from the Middle East. As per the statistics of the year 2017, Qatar released the highest emissions of about 49 tonnes per person, followed by Trinidad and Tobago (30 tonnes); Kuwait (25 tonnes); United Arab Emirates (25 tonnes); Brunei (24 tonnes); Bahrain (23 tonnes) and Saudi Arabia (19 tonnes) [11].

When we talk about the carbon footprint, the point shifts more toward the CO_2 emissions, which contribute 81% of the total GHGs. When looking at the CO_2 levels, these levels have risen to 440 ppm in the atmosphere which is maximum in 800,000 years. Annual carbon emission has reached a whopping 35 billion tonnes by 2019. As per the statistics, there has been an increase in the global average temperatures by more than 1 °C since preindustrial, which is even higher at the northern hemisphere with around 1.4 °C (by 2019). You may look at the sinister appearance of the statistics albeit the real situation is way more frightening [12, 15].

3 Environmental Concern and Regulations for Carbon Footprint

These carbon footprints are touching a new high yearly, which is becoming a significant concern for EPAs (Environmental Protection Agencies) and the scientific fraternity [16]. The question here arises; why higher carbon footprint is critical for the environment. Let us know about the associated problems.

Carbon footprints and the other GHGs absorb heat and work as a barrier, but increasing levels of these emissions are resulting in more heat absorption and a subsequent temperature rise, which has been considerably grown in the last 35 years [16]. Not only rising temperatures but it also has an impact on the environmental phenomenon such as thermal expansion, ocean acidification, and climate change. Rising temperature has imposed serious setbacks in climate change. Longer warmer seasons and shorter winter cycles with disturbed rain cycles are disturbing the crop cycle as well. Climate change imposes several potential challenges to the environment, physical and health impacts. These challenges are floods, severe droughts, and storms, heat waves, rising sea-levels, deviated crop growth; and disarrayed water systems. Furthermore, melting glaciers has also become a serious threat to rising sea levels. In recent years, as a result of thermal expansion and glacier melt, the global average sea level has been rising at about 3 mm per year during the past 20 years, threatening the inundation of areas with low elevation. Continuing a similar trend will submerge several islands in the India-Pacific Ocean regions in the coming future by 2050 to 2100 [17].

4 Probable Mitigation of These Concerns

A significant remedy to reduce these gigantic figures is very simple which is undo all activities that impelled us to reach here. Though it seems easy while talking but considerably impossible to undo which has been done for centuries. You cannot undo anything which has been taking place for years. This gives us two lessons; one is to try really hard to control as much as you can, and the other is to remember the limitation of "undo-rule." Presently, the world is way more concerned than ever before to convey as much as positive efforts. One commendable effort regarding the mitigation of climate change is to set a potential target by the UN member parties during the Paris Agreement. The aim was to limit the average global temperature to 2 °C above preindustrial temperatures, which are going to be discussed in this section [11].

Such huge targets may be easily achievable when deployed in certain industrial, transportation, and personal regulations. These are discussed as follows.

- Lesser dependency over luxury. Drive less, walk or cycle more, carpooling and mass transportation must help reduce emissions [11, 18]
- Choice of diet; preferring vegan diet over nonvegans; low carbon diets (7.19 kg-CO_2 equivalent for high meat-eaters and 2.89 for vegans) [19, 20]
- Follow the 4-R's formulae; Reduce, Reuse, Recycle, Refuse
- Use of lesser air-conditioning and heating. The natural systems have better air-conditioning and heating. This may be seen from the COVID-19 situation. Being dependent on natural sources is an important lesson taught by the current crisis.
- Reducing deforestation and increasing forestation and have proper care of that
- Utilization of renewable energy sources; biomass for metallurgical applications has proven itself and is a booming area of research. Not only biomass, other sources like solar, water and wind have also established in reducing the share of fossil fuels [7, 8]
- Reducing the use of fossil fuels as much as we can

- Industrial reforms for waste heat capture, recycling, and regeneration
- Mapping of industries with competent authorities
- Urban sustainable planning (such as sustainable cities)
- Nothing must go waste
- Technological upgrade
- Following the government protocols, schemes, and plans such as the Kyoto protocol, European Union—Emissions Trading Scheme (EU-ETS), Publicly Available Specification (PAS—2050), Paris agreement, etc. [21].
- Standardization of transportation mediums
- Return what you are taking from nature in whatever form you can
- Capital taxation over crossing the usable threshold for individuals as well as industries
- One child policy; In a study published by Environmental Research Letters in 2017 revealed the best way among all by following the one-child policy. Having one less child will help in mitigating our carbon footprint by an extensive amount. A reduction of 58.6 tonnes CO_2 equivalent may be observed in the case of developed countries [22].

These are some of the major reforms that the world needs to follow to capture the excess of these carbon emissions. Some of them can be followed by us very easily and others will be easily taken care of by the government and other competent agencies. Imagine a prosperous world, full of resources, clean air, water, land, ecological diversity, etc. with just small attempts by us. We are now aware of the carbon footprint, their effects on the environment, and mitigation methods. The next move is to understand what is their assessment and why it is needed which we are going to discuss in the next section.

5   Need for an Assessment

The actual quantitative understanding of the situation, which is being observed, is a result of the assessment. This is the reason why assessment is important. Assessment is nothing but a reasonable explanation and mapping of a process. It has nothing to do with the technological advancements, feasibility, and other processes/product attributes, just the mapping of the system. A complete portraying will come up with all the probable attributes of the system under consideration that gives detailed access to the system. The word system is given for a simpler term and that may be a process, product, transition, and approach. Assessment becomes necessary whenever you feel a requirement for system modification because then you may be able to precisely predict the possible outcomes. The assessment can be variable in terms of their types; qualitative and quantitative, but the objectives are going to serve the same. A typical example of an assessment method is carbon footprint assessment, and the results presented in previous sections are a form of quantitative assessment. Various tools may be used for this purpose but the fundamental principle is the same. The assessment becomes more useful when talking about the environmental impacts and carbon footprints and thus a robust assessment tool or approach is a mandate for a profound assessment result. Now the question that comes in our mind is how this assessment works? and how to identify its efficacy?

The detailed explanation that how the assessment works will be explained in the proceeding sections but for now, let us imagine a road trip as an example of assessment. There are a start and an end. When we said a road trip that means it is going to serve as the system, and start and end become the surroundings and limits for the system. Now next is the selection of parameters that must be selected on a most important basis. The reason behind this is to cover all the aspects most importantly the eminent ones that may have a significant impact on the system. When you have decided the parameter then starts the analysis that will take place while taking the road trip. As the trip ended you will have faced some occurrences that you have observed during the trip. At last, there comes the interpretation and recommendations, which are based on the analysis you have made during the trip which helps in identifying the problems, recommendations, and the other remarkable parameters that you may or may not miss. Now you have completed a part of the assessment that must be repeated in loops for a robust assessment. If you have successfully imagined the process then I guess, you may understand that we are doing such assessments on our regular basis. The more the system and the associated parameters will be complex, the more complicated is going to be the assessment. This is where the complex tools are applicable that helps in providing a detailed assessment which will be discussed at the end of this chapter.

An example of this type of assessment is the Personal Carbon Footprint. The method identifies the emissions released by an individual and that ultimately contributes to the total emission. The per capita emission values consist of the commodities apparently used by him in a specific period. Meanwhile, an efficient assessment will be one that will be comprised of all the possible aspects making the system profound.

There is an approach that has been successfully delivering assessment for a variety of applications and gaining importance worldwide and that is Life Cycle Assessment also known as LCA. As it is evident from the name LCA, it portrays the life cycle of a system. It is considerably famous for its role in decision supporting and environmental management, thus assessing the system from cradle to grave. The applications of LCA are limitless that we are going to find out as we will proceed further.

6 What Is LCA?

LCA is an environment management tool that assesses a process, product, and their probable outcomes from the cradle to grave stages [23–29]. In other words, LCA illustrates the sustainability and feasibility of the process and its objective under consideration. The process was assessed by recording and analyzing various parameters that result in a reasonable preparation of the inventory [30]. LCA can also be considered as a decision-supporting tool to evaluate a process in order to anticipate the positive and negative outcome of the system that whether it must be carried out or not [28, 31]. This makes it an essential tool in evaluating the system under consideration and its anticipated impact on the environment [28, 32].

LCA comprises of four steps, i.e., [25, 33]

(1) Goal and Scope
(2) LCA inventory generation
(3) Analysis and assessment
(4) Interpretation and discussion

6.1 Evolution of LCA and the Purpose of the Study [34]

The principles of LCA were developed in the 1960s. The LCA was initially named as Resource and Environmental Profile Analysis (REPA) by the US environment protection agency. In the earlier stage, it was more focused on the quantification of energy and resource consumption. The quantification of environmental impact was elementary work at that time. In the 1980s, assessments of the product life cycle experienced a renaissance through studies of the environmental loadings and potential impacts of beverage containers (e.g., beer cans, milk containers). This attempt started a parallel branch for the product life cycle assessment of the materials getting used in the packaging industry. In the late 1980s and 1990s, several worldwide attempts have been made to standardize the working mechanisms and its possible codes of conduct altogether to refine the assessment. This inclusion has grown the process so extensive that it started including very complex products such as paint, insulation materials, and kitchen and home appliances.

In 2000, LCA was standardized with a series of standards, i.e., 14000 series, which are illustrated in Table 1 [35].

Table 1 LCA stages with ISO standards [35, 36]	Standards	Stages of assessment
	Environmental Management 14040	LCA—Principles and Framework
	Environmental Management 14041	LCA—Goal and scope definition and Inventory Analysis
	Environmental Management 14042	LCA—Impact Assessment
	Environmental Management 14043	LCA—Interpretation
	Environmental Management 14044	LCA—Requirements and guidelines
	Environmental Management 14067	Greenhouse gases—Carbon footprint of products

ISO 14067 deals with the Greenhouse gases and the carbon footprint of products which assesses the emissions released during the lifecycle of the product. The life-cycle starts from the raw materials and ends at the "end-of-life" stage. The product-based LCA can be further divided into two parts that are; (a) cradle to gate, and (b) cradle to grave. Former measures the emissions from the raw material zone to the factory gate whereas the latter measure it from the raw material zone to the final disposal through product usage. Identification of the product LCA adhering to the standard can help you with finding the flaws in the system highlighting emission hotspots. Such types of assessments are equally fruitful for the manufacturer, clients, and, general public to obtain a benchmark against other products.

The product carbon labeling started in 2007, intending to let the users know about the footprint of the product being used.

Goals associated with the product carbon footprint

- To reduce the greenhouse gas emissions
- To identify the hotspots producing more and more emissions
- To manifest transparency to the product and their footprint
- To develop such products complying with international standards, regulations, and guidelines.

Since standardization in the last decade, LCA has become an environment management tool and also becomes an essential decision support tool in business, regulations, and policymaking, etc. LCA has also been involved in processes such as product and process improvement and design, environmental management, emission footprints auditing, resource management, best technology selection, etc. With this wide applicability, LCA can be considered as the sustainability assessment tool.

LCA is divided into four stages for assessing a product

Defining goal and scope: It includes defining the objective and the system boundaries of the product. The strategies for data collection and inventory generation and the evaluation methods are discussed in the stage. Additionally, an initial adjustment in the system boundaries and the parameters also takes place here.

Data Collection and treatment: Collection of data by measurements, calculations, literary evidence, peer reviews, the database search is performed in the stage.

Assessment: Incorporates normalization, classification, and assessment.

Interpretation: Involves sensitivity and reliability analysis and their assessment for technological feasibility.

6.2   Code for Impactful Conduct of LCA [36]

LCA is a significant and primary tool for the sustainability study and emerged with very wide applications from the last decades. Furthermore, LCA is a decision-supporting tool too. However, LCA also inherits some limitations and that is very certain. So, a code is present and that should be followed during the assessment to minimize the impact of the possible limitations. The code helps in conducting an impactful LCA. What are the code of conduct and the limitations and the ways it affects the study will be discussed in the further sections?

Data Quality: The quality of data involved in the study plays an important role in decision management by organizations. The data provided should not be kept on an as-received basis. A comparative study of the product with other products is necessary. The credibility of the study cannot be trusted if the assessment is done with the provided data from the manufacturers. Manipulation of data is possible and hence it must be validated with the secondary data, literary data, previous studies, and standardized data sets.

Life cycle boundaries: The system boundaries must be considered on its versatility basis. Not all the process boundaries can be tested and validated. Some processes do not synchronize with the assessment.

Country-specific technology types: The standard inventory articles are usually not applicable everywhere for conventional energy and renewable applications. The applicability of the processes and the incorporated technologies behave differently. So, the study and inventory management must be country specific.

Evaluation stage priorities: While comparing two products or processes, if the assessment depicts a similar kind of impact and the two systems are acquiring certain advantages over another. The ethical and rational decision is mandatory. Considering the present study, biomass replacement in metallurgical sintering is resulting in a positive reduction in carbon footprints and cost of the process but the technical feasibility is at stake. On the other hand, coal provides excellent technological feasibility but its emission generation behavior is not acceptable. So, a rational decision of combining both the fuels for technical feasibility and environmental acceptability can be a positive outcome and should be assessed further. The code is important because it speaks about the priority-based evaluation of the system.

Definition: Defining the system, the boundaries, the concerned objects, and the purpose must be clearly stated. Questions such as (a) why LCA is performed for the system, (b) how it will respond to the general structure, (c) how it will affect the user, and (d) how extensive are the study and the inventory involved. Without the understanding of these statements, An LCA will seem to be a directionless study.

Delimitation of the system under study: Once the definition is identified with a clear understanding, the system boundaries must be created carefully. System boundaries must be specific. A system under assessment can have numerous possible aspects, if the specific functionality is focused then the boundaries must be specific too.

Inventory: An inventory is the data set that provides the standard and the experimental data to assess the system. But the inventory must be answerable to certain questions such as (a) does the data includes all the process, (b) whether the data are collected from each stage or for a specific part, (c) the origin of the data source, (d) type of the data (standardized, literary, experimental, etc.), (e) quality of the data, (f) technological acceptability of the data, (e) impact of data on the environment, etc.

Impact assessment: The knowledge of the ways by which the system is impacting the environment or specified concerns must be specified. Whether the assessment is answering the response correctly and providing significant results is necessary. The motivation of the assessment is to understand the cradle to gate system and its impact on the concerning boundaries. A meaningful response for the desired study is expected and if it is available, it qualifies as a good conductor of LCA.

Evaluation: Evaluation must be answering the questions such as (a) are the assessment assessed all the objects carefully, (b) whether all the stages are checked or not, (c) transparency of the evaluation, (d) explanation of the response obtained, (e) what is your explanation towards the responses and (e) are you satisfied or not, if denying, then why and what changes required.

Learning objectives of LCA

The objective of the LCA study is the assessment of the impact on the environment. The impact enforced due to the unloading of the wastes and emissions and the loading of the resources is assessed here. The breakthroughs of the LCA are presented in a mere generalized way here:

- To explain the concept of the LCA and to study the impact created on the environment by the entire life cycle of the process and the product.
- To describe the process and the steps involved in a successful and impactful study.
- To reveal the conduct to follow LCA for better assessment results.

Types of LCA

LCA is divided into two types based on the factor that how a process or product is impacting the environment. Attributional life cycle assessment (ALCA) and consequential life cycle assessment (CLCA) are the two types of LCA [34]. Attributional LCA is the direct impact produced by the use of the product or process. It can be depicted as direct LCA. Consequential LCA explains the effect sustained by the environment during the process or the making of the product. CLCA is more of an assessment of the indirect impact. In the complete assessment of the process, it is essential to identify the type of LCA to be involved. Each LCA methodology is different from others and incorporates varied resources and frameworks. In the process modification and the biomass implication both the LCAs are involved. Where ALCA is associated with the cost of production study and also predicts the substantial impact on the environment by conventional coke usage and the biomass replacement. The ALCA will be defined with the characterization data and the input parameters cost study. CLCA is also a part of the study for assessing the impact caused by the biomass replacement in the modified sintering process and the comparative structure of the potential reduction in carbon footprint and other emissions footprint. CLCA will also depict the comparative response of biomass incorporation as an input parameter blended with conventional fuel.

Additionally, LCA is also divided into various types based on applications. It may be a process-based LCA, product-based LCA, and organizational LCA. The fundamental discourses among these types are the boundaries and scope whereas the methodology works the same.

7 LCA Methodology in Terms of Carbon Footprint Assessment

In the previous sections, LCA is introduced in general. As of now, we are quite aware of the working principle of LCA, methodology, conduct, and other parameters involved in it. Let us now talk about how this LCA takes place for carbon footprint assessment. In the upcoming sections, the detailed information on the stages of LCA pertaining to the carbon footprint is presented.

Each product or process has an impact on the environment. That impact may or may not be positive but that requires to be known as per the policies. In the case of carbon footprint, the same is not the result of an individual activity but it is a collective measure of human activities. These are some of the sectors that are contributing to these emissions extensively [24, 37–46].

- Iron and steel and other metallurgical operations
- Aluminum and other nonmetals production
- Power generation
- Mining
- Construction
- Municipal waste
- Windmills
- Fertilizer industry
- Hydropower
- Hydrogen generation
- Textiles
- Vehicles and other personal appliances
- Fuels, oil, petroleum
- Chemicals and paint
- Miscellaneous

Miscellaneous is dedicated to the sectors which are not listed here but are major contributors.

In the era of globalization, growth, race to become the economic superpower, increasing population, and lavish lifestyle, each country is looking for its resources to be utilized to maximize out of it. The necessity for expansion has been exhausting the resources for a really long time. Also, with the increasing demands of the goods and lifestyle has accelerated the production and other associated industries which has parallelly put up critical concerns for the environment. Years of the negligence of the environmental concerns have made us reach a point where we started facing issues that came into existence due to the negligence. These concerns are discussed earlier, one may refer it to previous sections. The expansion of human activities has expanded the carbon footprint up to the ferocious levels. We have reached a level where even if we stop releasing emissions from today, it will take around a decade to retrieve the wrongdoings. This is a theoretical proposition, which looks good to listen but restricting billion of tonnes of emissions to zero is impossible. But yes, we can integrate methods that will slowly lead to sustainability and help in reducing the emissions to some extent.

For the present context, we are going to assess the combined carbon footprint qualitatively. Here, the discussion will be more focused on carbon footprint assessment and therefore the discussion of the LCA stages will be comprised of such explanations. Individual assessment will be carried out in the upcoming sections where the real-time scenarios will be discussed in brief for each sector.

LCA is carried out in four stages. That is; defining goal and scope, inventory generation, impact assessment, and interpretation.

7.1 Goal and Scope

Defining the goal is a kind of task, which simultaneously acts as simple yet complex for the present investigation. One may think that how it is complex, but if we will start outlining the world's carbon footprint, it becomes really complex and time intensive to cover all the aspects. Leaving any aspect behind will put up some serious questions on the assessment and thus must be carried out with care. Let us understand this with an example. Suppose you are preparing for your final term examination but you are unaware of the syllabus. You can imagine the probable difficulties faced while preparing and the consequences, as well. Similarly, even when investigating small aspects, defining a goal and their scope becomes essential.

In the present context, the goal of carbon footprint assessment is to understand the insights into the full lifecycle of the processes, products, and organizations that are contributing to the frightening levels of carbon emissions. Usually, every activity imposes a footprint during their lifecycle which is called the activity's footprint, and as the context here is focused on the carbon emissions, it becomes an activity's carbon footprint. As the context is not activity specific thus the scope also inclines toward an extensive and nonspecific study or in other words a thorough assessment. The assessment is going to take into the account of the previously presented statistics as their datasets.

7.2 Inventory Generation

Now there comes the second stage of the LCA that is inventory preparation. Inventory is the most important factor of the LCA because it serves the ranges and limits of the campaign. For example, while discussing the footprint, how to tell the exact extent which is acceptable or not. In simpler form, inventory is something that gives quantification to the study. The inventory preparation is simpler while talking about a process or product because that is made up of some parameters. When the assessment is as big as the world's carbon footprint, the inventory generation becomes a huge and time-intensive task. However, when the goal and scopes are defined, inventory preparation becomes a bit relaxing.

I will present an example that will explain the statement. Suppose, you have given 10,000 currency (of any country), and you have to buy the groceries but you don't have the list of the things. Now you may feel some difficulty, even when you know that it is groceries that you have to buy. Similarly, the assessment here is very extensive and not activity specific and thus will comprise all the datasets pertaining to the carbon footprint that is presented in the previous sections. These datasets belong to the total carbon emissions being released into the environment from different countries and sectors. In other words, these datasets are going to serve as the inventory for the present study. The datasets represent the footprint of the individual product, process, individual's lifestyle, and the proposed moderations that are proposed in order to sustain the constraints being prioritized by the EPA's.

7.3 Impact Assessment

Now as we have acquired the inventory, the next step comes in is the impact assessment linked to the inventory. Let us explain this step with the help of a simpler example. If you remember the previous example which was about grocery shopping without a list. Now imagine, you accepted the challenge and went shopping with 10,000 bucks and without a list. What will happen? The only thing you will be able to do is randomly collect items that you seem necessary or the most essential ones but there are chances that the items are different from the expected ones. You may also face difficulty in selecting items to made them fit in the budget but you are going to fail to do so. These are the phenomena that happen with a faulty goal/scope/inventory, and these phenomena or occurrences are the impacts, and the stage assessing it is going to be called the impact assessment. Similarly, in the present context, we are going to discuss the impacts that are being advanced during the assessment or the implications made from this study.

Carbon footprint is an outcome of the human activities like globalization, expansion of industries, forest land acquisition, deforestation, excessive load on the resources, utilization of fossil fuels, open combustion, lavish lifestyle, dependency over high-carbon input technologies, absence of sustainability and greener mediums, time to time assessment and subsequent controlling, etc. The factors are boundless but that is something which is already done. Let us now discuss the consequences that we are facing at this moment.

The concentration of GHGs is the directly responsible factor for the rising earth's temperature. Higher the concentration, higher will be the earth's temperature. As the concentration is increasing the global earth's temperature is also rising. The rise in global temperature was 0.85 °C between 1885 and 2012, which has reached 1 °C in 2018 [16]. The reasons for the increased concentrations are already discussed in detail but a point worth noting is the use of fossil fuels. Around 2/3rd of these emissions are released directly from burning fossil fuels [17].

The impacts of the rising temperature on our atmosphere and environment are devastating. One of the most prominent effects is "Climate change." The impacts of climate change are resulting in catastrophic events such as disturbed weather patterns, deviation in the food production pattern, rising sea levels, melting of glaciers, etc. The impacts of climate change are global in scope and unprecedented in scale. As per the statistics, sea levels rose by 19 cm between 1901 and 2010, which will reach 24–30 cm by 2065, and 40–63 cm by 2100 [12].

7.4 Interpretation and Recommendation

Once the impacts are identified, the next comes the interpretation of these impacts, their remedies, and recommendation. The sinister impact of the increased carbon footprint is something that will require a globally collective attempt of human activities, governing bodies, regulatory, and protection agencies. If making a one-liner for the remedies and the recommendation, it must be "Undo the wrongdoing." It means a collective measure to regulate everything that is contributing to the footprint. Below are some of the most effective recommendations that will ensure a pause in the current rate.

The growing rate of industrialization has reached above par from the carbon footprint limits increasing the footprint remarkably. This expansion is a worrisome factor within the environment protection fraternity. As the demand is reaching the sky-high levels, the industries are bound to supply that for the growth of the country's infrastructure, and stopping it is going to deviate the overall order. The term industry is representative of all the industries in order to maintain a general homogeneity. The question that arises is "can this industrialization become eco-friendly"? or what are the methods to make it led toward sustainability in terms of environmental goals without deflecting their technological feasibility. It has been observed that integrating sustainability can deliver a more acceptable process in terms of the environment. The literature is full of these examples, which are presently under a lab-scale or batch-scale development or struggling to make space for inclusion.

The second major impact relies on the selection of material in various sectors. Every production industry starts with the preparation of raw material, and that raw material during their lifecycle reaches the final product for the use, and at last gets discarded for recycling, reuse, and disposal. The selection of a reasonable raw material is going to have a distinct impact on the environment on the positive side. The raw material if selected correctly will reduce the emissions and thus limiting the footprint. Next is the extended use of the goods or possibly reusing the product. This will release the stress from the resources and the manufacturing flow.

Another factor is the focus on the essential goods rather than the luxurious goods. This will also release the pressure from the industries. There is a famous saying that is relatable here is that "one must be firmed to return the resources that one is taking from nature." This saying can be possibly modified as "one must return the carbon footprint that one is imposing on the environment or if not return then one must save the equivalent." One major implication made is the reduce the dependency from conventional fuels like coal, coke, oil, and petroleum. These fuels and their subsidiary industries share more than half of the carbon footprint, which is a grave concern. These fossil or conventional fuels not only contribute to carbon footprint but also to other GHGs. Furthermore, the resources are getting exhausted with time with the current speed of consumption. To overcome that, the world is attempting to substitute them with renewables ensuring a low-carbon and cleaner methods of production, and now is the best time to accelerate the attempts for their potential utilization as the substitute.

As discussed earlier, we as humans are not behind as contributors. In order to procure a lavish lifestyle, we always forget that our existence is nothing without nature and it is not good to disturb the balance just for a temporary lifestyle. The use of appliances and commodities for every task is something that is disturbing the balance. These goods are made to serve us but right now we have extensively become dependent on them. Apart from the balance, these nonessential goods are expanding the carbon footprint largely.

The list of the implications and their corresponding impact is a very detailed topic to discuss and requires a dedicated chapter. Discussing it here will transform the assessment more into the philosophy. In the end, it may be concluded that the carbon footprints indeed impacting the environment badly and everything that is contributing must be controlled well. Also, unchecked and unregulated usage must be shut down or regulated to follow sustainability.

Now, as we have become familiar with the LCA methodology and their application in the total carbon footprint assessment to understand the impacts and their probable remedies. Let us discuss next the assessment of individual sectors contributing to the carbon footprint.

So, now we are going to assess the various sectors in terms of their carbon footprint, possible outcomes, and recommendations on how to deal with the expanding footprint.

8 Other LCA Examples

8.1 LCA of the Metallurgical Sintering Process for Carbon Footprint Assessment [27]

Previously, LCA is explained in a generalized way of assessing the carbon footprint for any general system. Let us now discuss the carbon footprint assessment for a complex process to understand the ground-level reality of the working mechanism of what we have learned previously. Before that, one should understand a bit about the sintering process. Sintering is a complex and intensive metallurgical process that provides the desired feed to iron and steel making operations. One may have some doubt that why a metallurgical operation is selected here. The main reason is the GHG contribution of the iron and steel sector. The statistics are huge for the emissions released by these metallurgical operations. Out of all, sintering contributes to a major proportion of GHG's. These contributions are due to the coke utilization in it which is added for heating and reducing purposes. It does work well for the assigned task but unknowingly contributes to emissions. Not only this but also the coal to coke carbonization also releases additional emissions. Let us see some statistics for sintering. The emission values released during coke making are not very significant but still noteworthy. These values were 0.001–1.230 kg/Mt for COx, 0.002–0.120 kg/Mt for SOx, and 0.09–3.173 kg/Mt for NOx (Source: Central Pollution Control Board (CPCB), Govt. of India). Meanwhile, a traditional sinter practice discharges around 241.53 kg of CO_2, 22.58 kg of CO, 0.294 kg of NOx, and 0.63 kg of SOx for per tonne of sinter [47]. Now imagine the emissions released for billions of tonnes of annual sinter production. To restrict that, a separate study carried out for utilizing biomass (sawdust and charcoal designated as SD and CH, respectively), that showed a reduction in those emission values released during conventional sintering. The upcoming sections will discuss how these modifications helped in the reduction of emission values. Additionally, LCA will also be used to compare the two to predict the feasibility of both processes, to follow up on the environmental goals (see Fig. 2).

Goal and Scope: The study here is about the assessment of sintering operation and its carbon footprint. Also, a comparative assessment of biomass and coke is manifested regarding the emission profile. Overall, the study outlines a conventional as well as a modified sintering process. The stage is further divided into four factors for a better distinction.

The functional unit: The investigatory campaign here revolves around metallurgical sintering operation. This operation is going to serve as the functional unit here. Also, there will be an additional modification in the fueling framework of the functional unit. A comparative functional unit for biomass and coke is illustrated in Fig. 1, which shows the variation in the fueling framework.

System boundaries: The assessment has been carried out to understand the effectiveness of the fuel substitution in terms of the carbon footprint as well as nitrogen

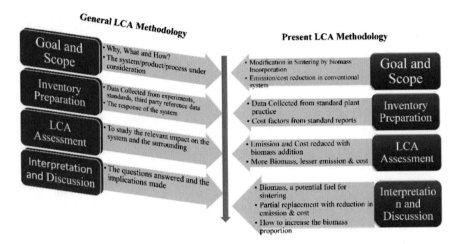

Fig. 2 LCA methodology

and sulfur footprint. It is tried to keep all the remaining factors unaltered to seek out a standardization, which will ensure zero deviation in operational parameters.

Data quality requirements: The datasets being collected are from the standard organizations, which are pioneers in their area of expertise. The datasets refer to the annual reports of IPCC and IBM (Intergovernmental Panel on Climate Change and Indian Bureau of Mines).

Impact categories: Elevated levels of GHG, waste utilization for energy, futuristic energy goals, and economic sustainability are the impacts that are addressed in the present study.

The present study manifested the following objectives. These are:

- To present a comparative assessment for the conventional and modified sintering process
- To check the emission profile for biomass substitution
- To acquire the reduction in emissive indices for coke substitution
- To highlight the other indirect factors that must be addressed well in future studies.

LCA Inventory Generation

As discussed earlier, the second and most crucial stage is the inventory generation, which is going to quantify the system boundaries and the system parameters. So, the data collection has been performed for the standard metallurgical sintering practice to interpret their emission profile and simultaneously carbon footprint. Later, the emission profile for the biomass-based sintering practice is calculated mathematically. These datasets will ensure a sustainable process following a cleaner production route if found environmentally acceptable. The dataset is indicative of the emissive indices released while producing one tonne of sinter. The carbon footprint for one tonne of sinter production is 220 kg CO_2 equivalent. Similarly, the same sinter production quantity releases 0.59 kg and 2.02 kg of NOx and SOx, respectively [48].

The emission profile mentioned in the inventory is referred and further stoichiometric calculations have been performed by mass balancing. As previously discussed, the carbon content of the fuel is solely responsible for generating the of COx emissions. In the experimental part of the study, coke had elemental carbon values of around 82.44%, which while reacting with the oxygen content of the iron-bearing phases, releases carbon emissions. If the carbon values are more, the discharge quantity will subsequently be higher. In the present case, coke's 82.44% carbon releases around 200 kg of CO_2 and 19.5 kg of CO, per tonne of traditional sinter produced. On the other hand, raw biomass possesses a relatively lower carbon proportion, which is around 45.73%. As the carbon proportions are low, their combustion will deliver a lesser carbon footprint. However, the sole usage of biomass for energy applications fails to provide the desired energy indices, still getting recommended for partial substitution. Hence, a partial substitution-based experimental campaign was followed in the study. Table 2 illustrates the emissive profile for successful experiments. One may observe the reduction in the emission values assisted with coke substitution.

As evident from Table 2, the coke-based sintering method releases the highest amount of about 222.11 kg emission per tonne of sinter. This carbon footprint sees a decline of 2.20% in the footprint indices when substituting 30% of coke with charcoal (See Table 3). Albeit the extent of reduction cannot be considered remarkable, any amount of decline in emission value is preferable. When looking at the response of coke substitution with saw dust, the trend acquired was quite similar. The best set of reduction was delivered with a combined fuel blend composition (CB:CH:SD::70:20:10), that obtained a total of 208.52 kg emissions for one tonne of

Table 2 Emission values for per tonne of sinter produced [27]

Process	CO_2 (kg)	CO (kg)	NOx (kg)	SOx (kg)	Total
Traditional Sintering	200	19.5	0.59	2.02	222.11
Modified Sintering *(70:20:10)	188.08	18.33	0.50	1.60	208.52
Modified Sintering **(70:30)	195.56	19.06	0.54	1.64	216.80
Modified Sintering ***(90:10)	190.92	18.60	0.54	1.85	211.92

*70:20:10—Coke (CB): Charcoal (CH): Sawdust (SD)
**70:30—Coke: Charcoal
***90: 10—Coke: Sawdust

Table 3 Reduction in Emission Values (%) [27]

Emission Reduction	CO_2 (%)	CO (%)	NOx (%)	SOx (%)
Modified Sintering (70:20:10)	6.00	6.00	15.25	20.79
Modified Sintering (70:30)	2.20	2.26	8.47	18.81
Modified Sintering (90:10)	4.54	4.62	8.47	8.42

sinter produced. This particular composition has succeeded in delivering a significant reduction for almost all the emissive constituents. NOx and SOx faced a noteworthy reduction among all (see Table 3).

8.1.1 Impact Assessment

As discussed earlier, the prime reason for the carbon footprint is fossil fuels like coal and coke [29]. The basic objective of using coal/coke for metallurgical operations is to make available the required energy profile for the desirable phenomenon and transitions during sintering. Some of these phenomena are fuel ignition, reduction, and partially melting the iron-bearing material and subsequently refining it by removing the oxygen proportion in it. The carbon present in the fuel reacts chemically with the oxygen present in the iron ore and reducing it to the next phase. During these events, there comes the generation of the COx, NOx, and SOx gases. The reduction and fusion occurring in the metallurgical processes can be partial or complete but the methodology remains the same.

These emissions released during sintering need to be captured before their release in the open environment, which is somewhat limited in the present times or not present everywhere. An example of these arrangements that are using these effluents are the integrated iron and steel plants. These industrial units use the gas, generated during blast furnace smelting, for sintering and in other processes for co-heating. But such uses are not very extensive. This is where the term like sustainability becomes important. The industrial practice currently is undoubtedly performing their assigned task well towards the technological goals but are we addressing the environmental acceptability appropriately? Leaving the environmental attribute makes the process vulnerable. When incorporating biomass, it was believed that the process will encounter that vulnerability precisely. Albeit the potential acceptable quantity is yet to be decided, it ensures a proactive reduction in emission profile and carbon footprint with very small substitution.

Interpretation and discussion

Biomass substitution in the fueling framework enables a significant reduction of the sintering's carbon footprint for an optimally attained biomass-coke blend under consideration, thereby contributing to future energy security. The optimal substitution is proportional to the fuel energy characteristics (under consideration). As we succeed in upgrading the energy and energy density values for biomass, then the reduction in carbon footprint will be a breakthrough in biomass applications. For now, interpretation of the assessment held out is discussed in a term-wise manner, which is as follows.

CO and CO$_2$ emissions: As previously discussed, COx emissions primitively originate from fuel, i.e., coke. When substituting coke with biomass, sintering gets optimized at a 30% substitution. The sintering quality indices obtained at this point were satisfactory with reduced emission profile as well as footprint.

NOx emissions: NOx emission gets introduced to the system from the air being sucked during combustion. Biomass integration as a substitution acquired a significant reduction of about 15.25% in the NOx values with just 30% of biomass substitution.

SOx emissions: SOx emissions released while sintering come from the sulfur present in the iron-bearing material and fossil fuel (coal/coke). The sulfur content present in the iron ore remains unchanged because it is not treatable even with a reasonable beneficiation practice, whereas sulfur removal from fuel also requires a chemical treatment that is presently being held for high sulfur coals. When not able to be treated, these sulfur present in the coke and iron ore results in SOx emissions during sintering. On the contrary, biomass contains 5–10 times less sulfur than coal, and using biomass will definitely lead to a reduction in sulfur emissions. During the experimental campaign, a significant reduction of 20.79% in SOx was achieved through a blended fuel composition of CB:CH:SD::70:20:10. This reduction was achieved with only a 30% substitution of coke. However, the substituted fractions are low and thus further reduction in sulfur emissions could be achieved in the future with higher proportions of biomass in the blend.

8.2 Carbon Footprint Assessment in Agriculture [49]

Previously, we learned the ways to assess the carbon footprint for metallurgical sintering operation. In the subsequent sections, we will discuss the carbon footprint assessment for other major areas apart from metallurgical operations.

Background: Agriculture is one of the major sources of the carbon footprint because of its potential to sequestrate carbon in the form of lignocellulosic biomass. Sequestration is basically the storage of carbon in the form of biomass by plants through photosynthesis. This sequestered carbon while burning or combusting mitigates that stored carbon releasing GHGs. As per the carbon storage potential, biomass contains around 30–40% of carbon, remaining is the moisture and volatiles. These compositions may vary with variable biomass sources. These carbon values were comparatively almost double in coke and coal. Also, the NOx and SOx values are 5–10 times lesser in biomass. When these constituents are lesser in biomass than coke then the released emissions will also be remarkably lesser. The emissions values will surely be discharged with biomass, but the extent will be relatively lower. This concludes an important aspect of biomass sources in industrial practice wherever coal or coke is the prime concern. Utilization of biomass will not only reduce GHGs, but this approach will also involve cleaner production which is the production of goods with cleaner means. This biomass substitution will be an easy option for the countries, which are rich in their agriculture sector considerably. Some of the agriculture rich countries are India, China, Sweden, etc. More stable agriculture will also produce agricultural wastes, which later on, with proper collection and preparation methods, resulting in solid biomass fuels. These fuels are presently being utilized in small-scale industries but they deserve rather better utilization considering their low emissive characteristic.

Objective: Carbon footprint assessment for biomass substitution.

Goal and Scope: The goal is to assess the reduction in the carbon footprint with or without biomass utilization. The context is fuel attributes. Fuel attributes are basically the fuel characteristics in terms of carbon footprint, carbon contents, energy, energy density, and cost of fuels.

Inventory generation: The next step is to generate the inventory for all the attributes that have been highlighted in the previous step. These data are from the experimental studies and standard industry practices. Table 4 displays a comparative scheme for biomass and coke that highlights the discourses.

Impact assessment: Higher carbon values present in the coke will be seen imposing a bigger carbon footprint than that of biomass. These high carbon values are likewise an indication of the high calorific values. Additionally, the energy densities are higher in coke than in biomass. The cost of production is higher for coke than biomass due to the nature of the fuel where coke is the processed product and biomass is raw in nature.

Interpretation and recommendation: High energy values provide better processing and thus desired for the industry but when comprehending the emission values, fails considerably. An opposite response is observed that signifies the potential of

Table 4 Comparative characteristics of coke and biomass [6]

Properties	Coke breeze	Charcoal	Saw dust
Moisture wt%	0.64	5.20	5.41
Ash wt%	20.30	9.63	8.81
Volatile matter wt%	2.88	20.78	69.79
Fixed carbon wt%	76.18	64.39	15.99
Carbon wt%	82.44	76.59	43.42
Hydrogen wt%	0.73	3.11	5.58
Nitrogen wt%	1.13	0.80	0.36
Sulfur wt%	0.78	0.61	0.09
Oxygen wt%	14.92	18.89	50.56
Al_2O_3 wt%	4.20	0.30	0.08
SiO_2 wt%	11.18	0.67	2.72
GCV (kcal/kg)	6024.10	5980.02	4421.67
Specific gravity (g/cm^3)	1.96	0.47	0.67
Cost (INR/tonne)	30000	12000	4000
Carbon footprint (kg CO_2 equivalent)	220.11	203.92	115.60

biomass. When discussing the production cost of these fuels, one must consider the processing methods for both the fuels (Coke and biomass) where the former requires high-temperature coke oven batteries, the desired quality coal through coal benefici- ation, storage, and transportation, and the latter requires a huge collection, storage, preparation, and transportation unit. Biomass does not require a high-temperature setup, which reduces the cost significantly.

The present assessment recommends the use of biomass over coke. The results are based on the environmental impact point of view. Albeit the energy limitations must be addressed well because biomass does possess lower energy values, which are questionable in terms of substitution potential.

8.3 Carbon Footprint Assessment for Aluminum Production [50]

Background: Next in line is the carbon footprint assessment for the aluminum production process. Alumina production is the next major contributing source of GHGs after the iron and steel sector. Alumina production potential is reaching sky high as the demands are increasing. As per the statistics, the global annual production of aluminum has reached 51 million tonnes in 2014 to 64 million tonnes in 2017, which will reach threefold by 2050. Russia, India, Canada, UAE, and Australia are the top 5 aluminum producers worldwide. One may understand the growing demand with the presented statistics which looks satisfactory with the point of view of infrastructure development and growth, but this will likewise have a negative impact on the environment due to the increased carbon footprint which is contributed from various subbing aluminum refining processes. Aluminum produc- tion discharges carbon emissions ranging between 7.2 and 44.5 kg (CO_2 equivalent per kg aluminum). Another aspect that is not generally introduced in previous studies is the red mud generation that is a huge environmental concern. Red mud is the waste product after leaching (hydrometallurgical processing), consisting of alumina, silica, iron, and titanium oxides. Presently, this red mud is a heated area on which studies are focused to separate the constituents of red mud for resource utilization and also to utilize these wastes. This red mud is not addressed in the previous peer reviews, which is an anticipated attribute.

Let us have a look at the attributes that are contributing to these values. A similar LCA approach will be followed as before.

Objective: Carbon footprint assessment for aluminum production

Goal and Scope: The goal of the study is to assess the carbon footprint produced during aluminum production. The study manifested various subbing processes to understand the individual contribution. The assessment starts from the ore mining to the final aluminum cast products to ensure a profoundness.

Table 5 An account of the carbon footprint from the aluminum production

Process	CO_2 (kg/tonne)	Global warming potential (kg CO_2 e/t)
Mining	22.9	23.5
Alumina refining	4682.4	4772.7
Anode production	506.6	512.1
Smelting	12,150.0	15497.5
Ingot casting	821.5	755.8

Inventory generation: The datasets are comprising of the emission data from individual processes from the mining to the final casting. A sample of the data is presented below in Table 5 [50].

Impact assessment: Carbon footprint is the largest for the alumina smelting among the subbing processes, followed by electrolytic refining of alumina. These carbon footprints are being contributed likewise from other processes. The main emitted greenhouse gases during the processes are CO_2, CH_4, C_2H_6, HFCs, PFCs, and SF_6. The emissions generated during alumina production accounts for 1% of the global carbon footprint. This may not look a huge amount when looking for the percentage but the reduction of a part of it will account for a significant reduction in the load being oppressed by the aluminum production.

Interpretation and recommendation: With the current industrial practice, the footprint is going to be enlarged with growing demands, which is clearly visible. Also, it is very difficult to reduce the requirements, then the only solution that is a mandate is the process/industrial sustainability. This will be done by approaches such as integrating emission collection, alternate methods for cleaner production, recycling, carbon trading, and other feasible technological methods. The assessment is prominent when the global industries are looking for a transition from the conventionalism to industry 4.0, mining 4.0, the industry of the future, sustainability toward technical, environmental, and economic goals.

8.4 Carbon Footprint from the Building and Construction Sector [44, 45]

Sustainable development has become a prominent concern for industries. Not only this but also the sustainable approaches such as green economy, sustainable planning, and management are gaining importance. Construction is one of the major sectors, which is directly interlinked to the infrastructural growth of any country. It is also one of the most dynamically growing industries. The construction growth is increasing with a growth rate of 5.2% per year with a total of 67% until 2020. China, USA, India, Japan, and Canada are the countries that are strongly contributing to this growth. In general, these five countries represent 50% of the total construction being carried out, worldwide. Looking at the growing pace of the construction sector, one can estimate the requirement of assessment. The context must be analyzed to study the probable environmental impacts to have an input–output mapping of the sector, especially for the high contributing countries.

The study intends to assess the carbon footprint for the global construction sector where the environmental impacts are caused due to the uncontrolled release of carbon emissions. Also, the accountable factors are highlighted to recommend imposing supervised control. The study conducted here is carried by Onat and Kucukvar [44], which presents a detailed analysis signifying the various subbing processes that contribute to the carbon footprint.

Objective: The carbon footprint of the global construction sector.

Goal and scope: The context here is the global construction sector where the unchecked outflow of carbon emissions is addressed. The study is focused on the five countries which are the highest contributors. These countries are China, USA, India, Japan, and Canada. The emissions are divided into three types; direct, indirect (regional), and indirect (global).

Inventory generation: Inventory is prepared from the standard databases named MRIO (multirange input–output), which represent the carbon emissions being released in each sector and for each country. A sample of the data is illustrated to provide the extent being released and the proportionate share of each country.

Impact assessment: China and India are the most populous country and thus growing their infrastructure significantly, which is also imperative of the carbon emission values shown in Table 6, having a share of 40% and 35%, respectively. However, there are limited studies on the sustainability studies as such, in which the present time demands extensively.

Interpretation and recommendation: With the increasing growth of countries and their economy, the construction is also growing multifold, which is also extending the carbon footprint. This extension of the footprint is unknowingly creating an impact on the environment and the results are clearly visible in recent times, which is getting worsen day by day. The remedy of this lies in the integration of sustainability to all the attributes. A comparative study of the construction sector with and without sustainable methods will ensure a transparent view of the reduction in shrinking the carbon footprint.

Table 6 Carbon footprint and their share [44]

Countries	Carbon emissions (Million tonnes)	Share (%)
China	1930	40
USA	376	8
India	1683	35
Japan	343	7
Canada	473	10

8.5 *Carbon Footprint Assessment of the Municipal Waste [51]*

Municipal waste is the everyday waste discarded from households, institutions, small businesses, office buildings, and other street and yard wastes. As per the report of Global Waste Management Outlook Report for the year 2015 (published by the International Solid Waste Association with the United Nations Environment Programme), these wastes generation increases annually with a rate of 2 billion tonnes/year. Municipal waste is mainly comprised of three types; construction wastes, residential waste, and industrial waste with a share of 36%, 24%, and 21%, respectively. These municipal wastes are either dumped or burned, openly which is a critical concern. Also, another critical concern arising is the capital investment for disposal and waste management. Here, the assessment has been carried out by Franca and team [51] where the study revolves around Rio de Janeiro which is the second-largest megacity of Brazil. The country produced around 78.6 million tonnes of waste in 2014. As per the statistics of Brazilian competent authority, in 2012, the city produced 22.6 million tonnes of CO_2, out of which 1.6 million tonnes were from urban waste disposal and 0.5 million tonnes from domestic and commercial sewage. A good and considerable way is to use these wastes as landfills but this is very selective in terms of wastes composition. Also, the objective must be more focused on recycle, reduce, and reutilize and then a controlled landfilling.

Here, an assessment has been carried out to provide the current trend of municipal waste in the city and their disposal scenario. Also, a trend of CO_2 generation is highlighted.

Objective: Carbon footprint assessment of the municipal waste generated in Rio de Janeiro.

Goal and Scope: The goal is to assess the carbon footprint of the municipal waste generated in Rio, and the efficient disposal in terms of management and utilization as controlled landfills.

Inventory generation: The datasets are collected from the city's municipal authorities. A sample of the data is displayed here that indicates the various sectors contributing to GHGs and their share (see Fig. 3). The data are based on Rio de Janeiro (Municipal Plan for Integrated Management of Solid Waste of Rio de Janeiro, 2012).

Impact assessment: Municipal waste is a day to day waste from our society, which when collected represents a huge amount. These wastes while burning generate GHGs and other effluents, which must be undone. Sustainable planning of the society and also proper waste utilization is the key factor in the efficient utilization of these wastes.

Interpretation and recommendation: The assessment recommends a sustainable utilization of municipal waste. Primarily, the proper collection, storage, and transportation strategy must be followed that will respond economically in the long run. Secondarily, the following pattern must be incorporated; nongeneration of wastes,

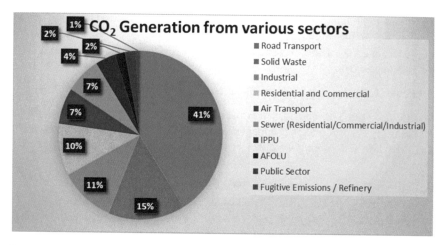

Fig. 3 CO_2 Generated from various sectors [51]

recycle, reduce, reutilize, and controlled landfilling of the remains. Controlled landfilling is important because the wastes may have an adverse effect on the soil and the underground water. One recommended way is their utilization in road construction.

8.6 Carbon Footprint Associated with the Textile Industry [52, 53]

All the previously discussed sectors including this are proportionally dependent on the economy, country growth, globalization, and population. Therefore, as these factors rise, the industry also expands, simultaneously expanding the carbon footprint.

Next in line is the textile industry, which is also a constantly expanding industry. Here, a product-based LCA was summarized. The lifecycle of a textile product is consisting of the following stages. These are as follows.

- Raw material preparation/processing/procurement
- Fiber and fabric production
- Garment production from these fabrics
- Packaging, distribution and retailing
- Consumer usage
- End life management, reuse, recycle, and/or disposal.

The study illustrated a lifecycle of a T-shirt that had a carbon footprint of 2.36 (CO_2 equivalent) throughout its life. This carbon footprint was further split up into two parts where 48% of the carbon footprint belonged to the "use phase," and the remaining 52% to the other phases, all combined. The 48% was mainly because of the ironing and tumble drying which may be substantially reduced if we will cut the ironing and tumble drying. Another important parameter is the type of fabric under consideration. 1 kg of cotton preparation emits 3.3 kg CO_2 whereas 1 kg of polyester (PET) releases 20 kg CO_2.

Objective: Carbon footprint assessment in the textile industry.

Goal and scope: The study assesses the carbon footprint from the textile industry, intending to seek out the environmental impact imposed by the textile industry. The study also aimed to highlight the factors, if avoided can result in shrinking of carbon footprint.

Inventory generation: LCA of the textiles often faces critical issues because of the unavailability of desired, full scale, dedicated, and standard datasets. This is due to the reasons that the work requires extensive work–power, time, and precision. However, some of the datasets do exist named Ecoinvent, GaBi, and ESU LCI database.

A sample dataset is displayed in Table 7 that illustrates the carbon footprint of a Levi Strauss & Co. Jeans produced for the American market in 2006.

Impact assessment: One can imagine the overall footprint when one cotton jeans have a footprint of 32.5 (kg CO_2 equivalent). Electricity is found as the most expensive commodity while producing any textile goods whereas the "use phase" is witnessed as the highly emitting zone of the product lifecycle. This carbon footprint can be reduced with certain steps that are listed in the next section.

Interpretation and recommendation: Following recommendations can be made from this assessment.

- Utilization of fabric that has the least footprint
- Use of renewables in manufacturing. Presently, everything is based on the electricity grid
- Use of sustainable technologies in the production
- Reuse and recycling
- Efficient disposal and incineration.
- Discarding the requirement for ironing and tumble-drying stage

Table 7 Carbon emissions in the production of one unit of jeans [53]

Stages	Carbon emission (kg CO_2 equivalent)
Cotton	1.7
Fabric production	6.6
Cut/sew/finish	3.0
Logistics and retailing	2.1
Consumer use	18.6
End of life	0.5

8.7 Carbon Footprint of a Windmill [38, 54, 55]

Renewable energy sources gained significance with time due to their low carbon economies and eco-friendly nature and ensured a low pollution environment and a lesser carbon footprint. One of them is wind energy that possesses remarkable energy potential. Wind energy is basically the energy generated from the conversion of wind's kinetic energy into mechanical power. The turbines convert the energy in the wind with the help of propellers.

Dependency over wind energy has witnessed drastic growth from the last two decades and it is the highest among all the other renewable energy sources. China, which is also the highest wind energy producer in the world, has grown its wind energy potential by 61.8% during 2004 and 2013 (US DOE 2014). China has succeeded in installing a total capacity of 221 GW by 2019, followed by the USA and Germany with 96.4 GW and 59.3 GW.

These sources are undoubtedly significant renewable sources, Albeit, unwillingly contributes to the carbon footprint. As per the statistics, in China, a windmill generates 0.15 million tonnes of CO_2 in its lifetime of 21 years. The majority is due to the metal and steel construction of the mills. The major contributing sources are construction, operation, and dismantling with a share of 76.74%, 15.32%, and 7.94%, respectively.

Objective: To assess the carbon footprint associated with the lifecycle of a windmill.

Goal and scope: The study focuses on the windmill lifecycle that how the lifecycle is contributing to the footprint and their distribution.

Inventory generation: Inventory shows the share of carbon emission being released during the entire lifecycle of windmill operation. Life is typically considered around 21 years. The construction phase of the windmill imposes the highest carbon footprint among other phases such as operation and dismantling (see Table 8).

Impact assessment: Wind energy is a proven and potential source of energy that drives the shift of the dependency from fossil fuels to the renewables. However, it generates some emissions during the process. The emissions are not direct ones and can be recorded as indirect ones. The footprint is manageable when looking at their renewable energy outcome, still, these emissions must not be overlooked.

Table 8 Carbon emissions from the various sources of windmill construction [38]

Sources	Share (%)
Smelting and processing of metals	44.72
Scientific Research	38.96
Construction	22.04
Electricity and heating power production and supply	21.33
Electric equipment and machinery	20.00
Nonmetal mineral products	7.73

Interpretation and recommendation: If for the instance, we forget this indirect footprint, wind energy is an energy source for the future. The energy is not coming at any cost, except for the extras that are essential for the construction. One may look at the ways to reduce these footprints with the help of sustainable technologies and replacement of metals with other materials that can be prepared with low carbon mediums.

These LCA examples were manifested in order to provide insight and understanding to the readers. LCA can be integrated with ease to the places where ever the environment is a critical concern. In the next sections, we will discuss the tools that are currently being used for the LCA. These tools will serve more efficiently when the study is related to any organization or when there is a large set of parameters and datasets.

9 Tools and Software's Used for LCA

LCA seems to be an easy task when you are assessing a product with a limited number of parameters but the task becomes complex and difficult when you need to assess a process that has a large number of parameters or while assessing the carbon footprint of any organization [56]. To overcome these difficulties, the use of software tools becomes interesting. These tools are readied with compliance with the standards and protocols, hence delivers a transparent assessment. Even when your organization is dealing with a variety of products, these tools will address altogether with ease. Some of them are listed below with their features and applicability. Some of them may be missed but the aim was to discover the way how this software works.

9.1 Air.e HdC

Events like climate change are progressing rapidly, which is something inevitable, the only way to reduce it is by lowering down the carbon emission levels. The first step in line is to quantify the footprint. This is where this software comes in. Air.e HdC is a software solution used by many industries to analyze and calculate the carbon footprint and GHG emissions for their product, process, and services. The software works in compliance with all the essential standardization and protocols such as ISO 14064, ISO 14067, GHG Protocol, and PAS 2050 [57].

Additionally, this software package comes with the following features, which are:

- Powerful user-friendly interface
- Added features such as LCA versions management, nested LCA, co-products, carbon sinks, import and exports
- Powerful relational database and searching facility
- Frequent upgrade of the database and real-time calculation

9.2   Ecodesign+

The software helps in analyzing the product's carbon footprint and its mapping with the product in order to improve the footprint. The software is easy to use, intuitive, and does not require expertise in environmental assessments. ECODESIGN+ uses LCI database Ecoinvent (v2.2) and established a full life cycle assessment (LCA) practices (ISO 14040/44) for its calculations [58].

Software's salient features:

- A tool to calculate, improve, understand, interconnection, and mapping
- Intuitive tool for designers, developers, and technical experts
- Quick reporting and faster results generation
- Standard database and compliance
- Practical teaching tool
- Runs on any browser and easy to use interface

9.3   CleanMetrics

The software tool helps to calculate the carbon, water, and energy footprint with a standard based methodology. It takes into account all the critical concerns of the product, process, and services [59]. These are

- Production, transportation of the materials, components and other supplies
- Carbon sequestration and mitigation
- Lifecycle carbon sequestration
- Time-bound GHG assessment
- End of life or disposal of product and service

The software complies with all the required standards and a huge database. Additionally, the tool has wide applicability in areas such as energy, construction, textiles, transportation, food, and furniture.

9.4   Clean CO_2

This LCA software is loaded with features that enable the easy calculation and analysis of the carbon footprint of any organization. It provides ease in identifying the footprint of the overall organization. This tool is presently being used by many companies to satisfactorily offset their carbon footprint which will have a positive impact on the environment. The software complies with all the standards and protocols such as ISO 14067, PAS 2050, GHG Protocol Product Accounting and Reporting [60].

9.5 SimaPro

It is one of the leading software tools widely being used since 2003, among the LCA specialists. It has one of the largest databases such as ecoinvent, industry-specific Agri-footprint database, and the ELCD database, U.S. Life Cycle Inventory Database (USLCI), Danish Input-Output Database 99, etc. The software becomes more efficient when you feed a transparent, real, and high-quality inventory data [61].

9.6 One-Click LCA

One-click LCA is a software tool that automates the assessment task to calculate the carbon footprint and other environmental impacts.

10 Summary

The chapter introduced us about terms like carbon footprint, assessment, LCA, their application for various areas. Additionally, the chapter possibly delivered the answers of questions such as why assessment is needed, why LCA, how LCA works, and all the associated principles. Carbon footprint assessment is a very common understanding nowadays but when you dig deeper to find their reach and impacts, you will find out that our environment is getting significantly affected by these footprints. Moreover, these emissions are not just increasing in numbers, it is rising serious threats to the environment as well as humanity by alleviating earth's atmospheric temperature, which causes climate change. In a nutshell, our activities have disturbed the overall system for which the world is looking for answers. The chapter has also highlighted some key issues that are the result of increased carbon footprint and the collective measure to reduce the footprint.

LCA is merely a tool or approach to calculate and analyze the carbon footprint of any product, process, system, organization, service, etc. It is more likely an environmental management and decision support tool. LCA is presently being used for almost every application for other footprints as well such as water, food, energy, etc. In this chapter, the LCA approach is used to qualitatively assess the overall earth's carbon footprint, the comparative footprint of biomass (renewable) and coke (fossil fuel), and others.

Acknowledgements The authors are thankful to the researchers, which are referred to in the article for their indirect contribution to this chapter and direct contribution to the carbon footprint assessment. It is really difficult to name each one but, I believe a collective acknowledgment will be appreciated.

References

1. What Is a Carbon Footprint?
2. Wright LA, Kemp S, Williams I (2011) Carbon Manag, 2:2
3. P. Association: Guard
4. Jones CM, Kammen DM (2011) Environ Sci Technol 45:4088–4095
5. Online Supporting Data, Calculations & Methodologies for Paper: Jones, Kammen 'Quantifying Carbon Footprint Reduction Opportunities for U.S. Households and Communities' ES&T, 2011 (Publicly Available)
6. Jha G, Soren S, Deo K (2020) Mehta: fuel 278:118350
7. Jha G, Soren S, Renew Sustain Energy Rev. https://doi.org/10.1016/j.rser.2017.05.246
8. Klass DL (1998) Biomass Renew Energy Fuels Chem, 137–158
9. Zandi M, Martinez-Pacheco M, Fray TAT (2010) Miner Eng 23:1139–1145
10. Fröhlichová M, Legemza J, Findorák R , Mašlejová A, Arch Metall Mater. https://doi.org/10.2478/amm-2014-0139
11. Carbon footprint—Wikipedia, https://en.wikipedia.org/wiki/Carbon_footprint. Accessed 29 July 2020
12. Our World Data
13. CO_2 and Greenhouse Gas Emissions—Our World in Data, https://ourworldindata.org/co2-and-other-greenhouse-gas-emissions. Accessed 29 July 2020
14. Climate Change & the Carbon Footprint—Global Footprint Network, https://www.footprintnetwork.org/our-work/climate-change/. Accessed 29 July 2020
15. Ritchie H, Roser M, Annual total CO_2 emissions, by world region
16. O. US EPA: US EPA
17. The Cost of Energy, Environmental Impact—The National Academies, http://needtoknow.nas.edu/energy/energy-costs/environmental/. Accessed 29 July 2020
18. R. Harrabin: BBC
19. Scarborough P, Appleby PN, Mizdrak A, Briggs ADM, Travis RC, Bradbury KE, Key TJ (2014) Clim Change 125:179–192
20. Eshel G, Martin PA (2006) Earth Interact 10:1–17
21. bsigroup.com
22. Weindl IS, Lotze-Campen H, Popp A, Müller C, Havlík P, Herrero M, Schmitz C, Weindl RI, Lotze-Campen H, Popp A, Müller C, Havlík P, Herrero M, Schmitz C, Rolinski S (2015) Environ Res Lett 10:094021
23. Chisalita DA, Petrescu L, Cobden P, van Dijk HAJE, Cormos AM, Cormos CC (2019) J Clean Prod 211:1015–25
24. Keller F, Lee RP, Meyer B (2019) J Clean Prod 250:119484
25. Laurin L, Global E, States U (2017) Overview of LCA d history, concept, and methodology, vol 1, Elsevier
26. Schmidt A, Poulsen PB, Andreasen J, Fløe T, Poulsen KE, C.B. a S: 2004, p 180
27. Jha G, Soren S, Mehta KD (2020) J Clean Prod 259:120889
28. Norgate T, Haque N (2013) Miner Eng 42:13–21
29. Tomkins P, Müller TE, Green Chem. https://doi.org/10.1039/c9gc00528e
30. Heijungs R, Hellweg S, Koehler A, Pennington D, Suh S (2009) J Environ Manage 91:1–21
31. Rahman SMM, Handler RM, Mayer AL, J Clean Prod. https://doi.org/10.1016/j.jclepro.2016.07.014
32. Adiansyah JS, Rosano M, Biswas W, Haque N (2017) J Sustain Min 16:114–125
33. Fi LO, Thonemann N, Pizzol M, Khandelwal H, Thalla AK, Kumar S, Kumar R (2019) Bioresour Technol, 121515
34. C. 3: 2008, pp 41–9
35. International Organization for Standardization (2006) Int Organ Stand 3:20
36. Int. Organ. Stand
37. Chojnacka K, Kowalski Z, Kulczycka J, Dmytryk A, Górecki H, Ligas B, Gramza M (2019) J Environ Manage 231:962–967

38. Ji S, Chen B (2016) Energy Procedia 88:250–256
39. Kylili A, Christoforou E, Fokaides PA, Biomass and Bioenergy. https://doi.org/10.1016/j.bio mbioe.2015.11.018
40. Pasqualino J, Meneses M, Castells F (2011) J Food Eng 103:357–365
41. Bartzas G, Zaharaki D, Komnitsas K (2015) Inf Process Agric 2:191–207
42. Guo Y, Zhu W, Yang Y, Cheng H (2019) J Clean Prod 239:118004
43. Von Der Assen N, Voll P, Peters M, Bardow A (2014) Chem Soc Rev 43:7982–7994
44. Cihat N, Kucukvar M (2020) Renew Sustain Energy Rev 124:109783
45. Li X, Zheng Y (2019) J Clean Prod, 118754
46. Meng F, Ibbett R, de Vrije T, Metcalf P, Tucker G, McKechnie J (2019) Waste Manag 89:177–189
47. Fortier MP, Teron L, Reames TG, Trishana D, Sullivan BM (2019) Appl Energy 236:211–219
48. Fi LO. https://doi.org/10.1016/s1006-706x(15)30029-7
49. Adewale C, Reganold JP, Higgins S, Evans RD, Carpenter-boggs L (2019) J Clean Prod 229:795–805
50. Gautam M, Pandey B, Agrawal M (2016) carbon footprint of aluminum production : emissions and mitigation, Elsevier Inc
51. Franca LS, Rocha MSR, Ribeiro GM (2018) carbon footprint of municipal solid waste considering selective collection of recyclable waste, Elsevier Inc
52. https://doi.org/10.1016/b978-0-12-819783-7.00007-7
53. Impact E, Chain CS, https://doi.org/10.1016/b978-0-12-819783-7.00009-0
54. Power SW (2014), p 16
55. Thompson P, Anderson DR, Fisher R, Thompson D, Sharp JH (2003) Fuel 82:2125–2137
56. Rattanatum T, Frauzem R, Malakul P, Gani R (2018) LCSoft as a Tool for LCA : New LCIA methodologies and interpretation, vol. 43, Elsevier Masson SAS
57. Air.e HdC—Carbon Footprint Software by Solid Forest… https://www.environmental-expert. com/software/air-e-hdc-carbon-footprint-software-514749. Accessed 29 July 2020
58. Ecodesign+—Product Carbon Footprint made easy. http://esu-services.ch/software/ecodesign plus/. Accessed 29 July 2020
59. Product LCA and Carbon Footprint Analysis - CleanMetrics 2.0. https://www.cleanmetrics. com/CarbonFootprints. Accessed 29 July 2020
60. Clean CO2 -Calculating and offsetting emissions of organizations, products or services, http:// esu-services.ch/software/cleanco2/. Accessed 29 July 2020
61. SimaPro Software—The regional SimaPro competence centre for Switzerland, Austria, Germany and Liechtenstein. http://esu-services.ch/simapro/. Accessed 29 July 2020

Evaluating Impacts of Traffic Incidents on CO_2 Emissions in Express Roads

Marina Leite de Barros Baltar, Victor Hugo Souza de Abreu, Glaydston Mattos Ribeiro, and Andrea Souza Santos

Abstract Traffic congestion is common in large cities and on important express roads, which imposes travel time delays, an excessive fossil fuel consumption, an increased environmental pollution, etc. Related to it, traffic incidents such as broken-down vehicles, accidents and flat tires intensify the traffic congestion because they generate irregular but frequent interruptions. Thus, it is necessary to implement studies that seek to minimize the impacts of incidents. Therefore, this chapter seeks to apply the MEET model (Methodologies for Estimating air pollutant Emissions from Transport) to analyze the impact of incidents on carbon dioxide (CO_2) emissions on express roads. For conducting the case study tests, real data are used from approximately 2,800 incidents on Avenida Brasil, the main expressway in the Rio de Janeiro city, which were provided by the Traffic Engineering Company of the Rio de Janeiro city (CET-Rio). The results indicate that incidents increase in 22% the CO_2 emissions, that the broken-down vehicles are the incidents with the greatest impact on these emissions due to their high frequencies, and that the morning and afternoon peak hours are responsible for 82.4% of the increase in CO_2 emissions related to the incidents. In addition, using Kernel maps, it is possible to verify the sections with the highest incidents, as well as CO_2 emissions.

Keywords Traffic management · Traffic incidents · MEET model · Environmental pollution · Carbon dioxide · CO_2 emissions · Express roads · Rio de janeiro

M. L. de Barros Baltar · V. H. S. de Abreu · G. M. Ribeiro (✉) · A. S. Santos
Transport Engineering Program (PET), Alberto Luiz Coimbra Institute for Graduate Studies and Research in Engineering (COPPE), Federal University of Rio de Janeiro (UFRJ), Rio de Janeiro, Brazil
e-mail: glaydston@pet.coppe.ufrj.br

M. L. de Barros Baltar
e-mail: mabaltar@pet.coppe.ufrj.br

V. H. S. de Abreu
e-mail: victor@pet.coppe.ufrj.br

A. S. Santos
e-mail: andrea.santos@pet.coppe.ufrj.br

1   Introduction

The high degree of urbanization and the economic growth of the cities, although opening the way for countless achievements, also introduced challenges such as traffic congestion, which is one of the major urban problems faced in recent decades [4]. The reason is that congestion generates road interruptions that cause travel time delays [15], excessive fossil fuel consumption and increased environmental pollution [22, 33].

Congestion can be intensified by events such as traffic incidents (broken-down vehicles, accidents, flat tires and others), which generate irregular but frequent road interruptions, as a result of the intermittent traffic flow, generated by the condition imposed by bottlenecks on the runway [6, 13, 37, 42].

To overcome these problems, a new transport infrastructure must be designed and built, or traffic policies must be developed to manage and control the traffic flow. The latter is advantageous mainly in a sustainability point of view because it exploits the existing infrastructure [21], avoiding unnecessary expenses which are almost always infeasible due to budget restrictions of the management and control agencies.

Thus, traffic management and control strategies are widely applied to mitigate congestion as they seek to smooth traffic flows, reduce travel time and minimize pollutant emissions [10]. Specifically, in relation to vehicle pollutant emissions, including greenhouse gases (GHG) and pollutants harmful to air quality, urban areas produce a disproportionate amount compared with their geographic size. Thus, local authorities and transport decision-makers strongly need to mitigate these emissions [22].

In this context, this chapter aims to evaluate, through the application of the MEET model (Methodologies for Estimating air pollutant Emissions from Transport), the impacts on carbon dioxide (CO_2) emissions on express roads caused by traffic incidents, which are related to the speed reduction. Rio de Janeiro city is used as a case study to evaluate these emissions and to propose mitigation actions. The scope of this study is limited to CO_2, considered the major contributor to GHG emissions during the transportation tasks [22].

In addition, we know that integrated strategies to reduce GHG and air pollutant emissions result in significant cobenefits [29, 38], such as improving air quality and reducing public health spending. Therefore, the overall effect of a transport intervention (or combined effect of many small and diffuse interventions) on CO_2 emissions is more important than any localized effects [22].

In order to verify the impacts, real data are used from approximately 2,800 incidents on Avenida Brasil, the main expressway in the Rio de Janeiro city. These data were provided by the Traffic Engineering Company of the Rio de Janeiro city (*Companhia de Engenharia de Tráfego da cidade do Rio de Janeiro*—CET-Rio, in portuguese). This database contains the location of the incident, its time and the vehicle involved.

We use the MEET model [25] to determine the CO_2 emissions because, according to Bai et al. [4]: (i) it is based on measurements on the road, so the parameters are extracted from real-life experiences; (ii) it is sensitive to speed, which is an important reference and easy to obtain; (iii) it has a wide coverage of possible combinations of technologies and fuels; and (iv) it considers many factors such as fuel types, engines and vehicle types.

In order to achieve the objectives of this study, the remainder of this chapter is organized as follows. Section 2 presents the methodology used. Section 3 describes the MEET model needed to estimate CO_2 emissions considering two cases: with and without incidents. Section 4 describes the case study, presents and discusses the results and indicates the possible actions necessary to minimize the problems encountered. Finally, Sect. 5 presents the final considerations of this study.

2 Methodology

Traffic incidents contribute to the increase in CO_2 emissions due to the congestion generated, however efficient and quick responses to the incidents lead to the reduction of this externality [13]. In addition, environmental, social and political pressures to reduce the impacts associated with CO_2 emissions are increasing rapidly [18]. Thus, it becomes necessary to develop strategies to minimize these problems [22].

Therefore, the methodology proposed in this study is divided into six stages, as shown in Fig. 1, which seeks to: (i) estimate the increase in CO_2 emissions, due to the occurrence of incidents; (ii) analyze which incident type most aggravates the situation; and (iii) propose actions in order to mitigate the problem.

In Stage 1, a bibliographic review is performed in order to identify studies about the impact of emissions on the urban environment, as well as ways to estimate emissions and the intrinsic relationship between CO_2 emissions and traffic speed. In Stage 2, the mathematical model needed to calculate CO_2 emissions on an urban road is determined, based on the speed reduction caused by the incidents.

In Stage 3, data are collected such as the history of traffic incidents, the location of the occurrences, the flow parameters (vehicle speed and flow) and the traffic composition, i.e., the percentage of cars, buses and trucks. In Stage 4, CO_2 emissions are estimated considering the occurrences of incidents by applying the MEET model. It is worth mentioning that the duration of the incident is considered from the moment that the vehicle impedes the regular traffic flow until the moment when it is removed. This stage also estimates the CO_2 emissions without incidents (considering only the regular traffic flow). To compare the CO_2 emissions with and without the incident, we considered the same duration time (the duration of the incident).

In Stage 5, an exploratory analysis of the results is performed. The analysis is performed globally, seeking to evaluate the impact on CO_2 emissions due to the occurrence of incidents, and then it is carried out in a grouped manner, seeking to understand the incident type, time and vehicle that most generates emissions. In Stage 6, there is a discussion and proposal for actions that can be taken to reduce CO_2 emissions, based on the analysis of the previous Stages 1 and 5.

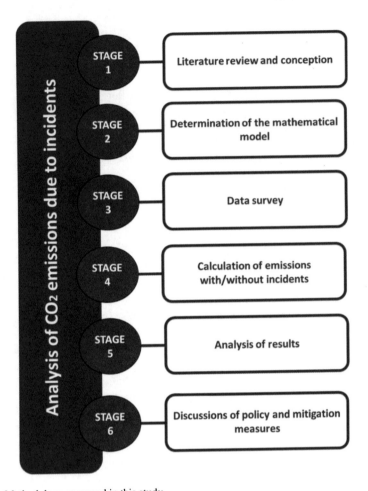

Fig. 1 Methodology proposed in this study

3 Model to Estimate CO_2 Emissions

To estimate CO_2 emissions, we use the MEET model [25], which relies on real measurements to estimate transport emissions and energy consumptions for different vehicle types taking into account the speed reduction. This relationship is also studied by other authors like Smit et al. [36] and Abou-Senna and Radwan [1].

The calculation of emissions in grams per kilometer ($E(v)$) is based on the vehicle types under analysis and on the speed value (v), as presented in Eq. (1).

$$E(v) = K + av + bv^2 + cv^3 + d/v + e/v^2 + f/v^3 \qquad (1)$$

The values of the constants K, a, b, c, d, e and f vary according to the vehicle type under analysis and they are obtained from measurements performed considering real cases [25]. Three vehicle types were considered for the coefficients of Eq. (1), as proposed by Bai et al. [4]: (i) vehicles smaller than 3.5 tons, representing passenger cars and small cargo vehicles; (ii) vehicles between 3.5 and 7.5 tons, which represent trucks; and (iii) vehicles between 3.5 and 16 tons, which are buses, as shown in Fig. 2. These vehicle types represent the ones found in Avenida Brasil.

So, Table 1 shows the values used for the constants presented in Eq. (1) for each vehicle type. These values were proposed by Hickman et al. [25].

Although MEET model allows calculating different emissions types, it was decided to focus the present study on CO_2 emissions, since the increase in the concentration of this gas in the atmosphere has led to serious environmental problems such as global warming and climate change, as demonstrated by Grote et al. [22], Jabali et al. [27], DECC [14] and IPCC [26].

Initially, to estimate the total CO_2 emission, it is necessary to define the number of vehicles affected by the incidents by surveying the average traffic flow that would be passing on the road at the time of the incident. However, depending on

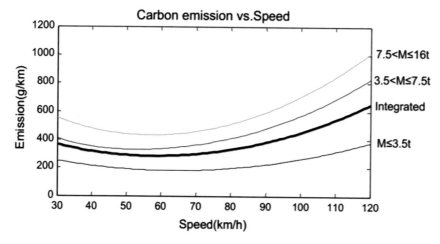

Fig. 2 Carbon emission by vehicle type. *Source* Bai et al. [4]

Table 1 Values for the constants used in Eq. (1)

Vehicle type	K	a	b	c	d	e	f
Passenger cars and small cargo vehicles	601.2	−9.8381	0.0621	0	0	0	0
Trucks	110	0	0	0.000375	8702	0	0
Buses	871	−16	0.143	0	0	32031	−

Source Hickman et al. [25]

the traffic conditions of the road, when an incident occurs, part of the flow overcomes the bottleneck while another part is retained, forming a queue of non-recurring congestion.

Therefore, it was necessary to define how many vehicles passed through the incident and how many stayed in the traffic congestion queue (q). So, let f be the regular traffic flow at the time of the incident and C be the road capacity that is reduced in the section under study during the incident's time. If $f \leq C$, there is no nonrecurrent congestion due an incident, but the speed can be reduced. In this scenario, when there is an incident, the emission calculation is performed according to Eq. (2). The values of e_{pc}, e_t and e_b are obtained using Eq. (1) with their correspondent constants according to vehicle types.

$$E_{tot1} = t \times f \times \left(\% f_{pc} \times e_{pc} + \% f_t \times e_t + \% f_b \times e_b \right) \tag{2}$$

where:

- t is the time duration of the incident;
- $\% f_{pc}$ is the percentage of the passenger cars in the traffic composition;
- e_{pc} is the passenger car emission rate;
- $\% f_t$ is the percentage of the trucks in the traffic composition;
- e_t is the truck emission rate;
- $\% f_b$ is the percentage of the bus in the traffic composition; and
- e_b is the bus emission rate.

However, if $f > C$, there are traffic congestion queue in such way that the size (q) can be estimated from the Queuing Theory [19], according to Eq. (3). The result of this equation is the number of vehicles in queue.

$$q = |ft| - \left| \left(t - \frac{1}{f} \right) C \right| \tag{3}$$

For this case, the total emission will be calculated according to Eq. (4). The parameters are the ones presented before.

$$E_{tot2} = q \times \left(\% f_{pc} \times e_{pc} + \% f_t \times e_t + \% f_b \times e_b \right)$$
$$+ t \times C \times \left(\% f_{pc} \times e_{pc} + \% f_t \times e_t + \% f_b \times e_b \right) \tag{4}$$

Equations (2) and (4) seek to estimate the CO_2 emissions considering the occurrences of incidents. The first equation is used if the incident does not increase the recurrent queue and the second one is used if a queue is generated by the capacity reduction at incidents.

To understand the difference in CO_2 emissions with and without incidents, it is necessary to compare the two scenarios, as proposed in Stage 4 of the methodology presented in Sect. 2. As the total emission increases over the time (the higher the duration considerer, higher is the emission), was considered the same duration time for the two cases. So, to estimate the CO_2 emissions without incidents, we used Eq. (2). The same traffic flow of the scenario with incidents was considered, but the capacity and speed reduction imposed by the incidents were not.

In summary, Eq. (2) is used for scenarios with incidents that do not generate additional congestion queues and for the scenario without incidents, only the speed values change in these two cases. Finally, Eq. (4) is used in scenarios with incidents that generate congestion queues.

4 Case Study

Traffic congestion aggravates air pollutant emissions in urban areas [22]. In addition to recurrent congestion in cities which occur due to lack of capacity in the system, there are nonrecurring congestions due to incidents that aggravate the situation of urban mobility [37, 42].

In the Rio de Janeiro city, this reality is not different. The city suffers of daily traffic congestion that is intensified by the incidents, which have serious social, economic and environmental impacts. So, in order to reduce these problems, CET-Rio performs road operations for service the incidents using tows trucks, motorcycles and pickups. Considering the tows trucks, CET-Rio has three types: (i) light tow truck for passenger cars; (ii) heavy tow truck for trucks and buses; and (iii) super heavy tows trucks for big trucks.

In order to verify the impacts of the incidents, real data were obtained in the Rio de Janeiro city. One of the main express roads is Avenida Brasil, which is the largest road in the city with 58 km and with an expressive number of calls to attend incidents. It is the main road connecting the city center, the suburb and the west zone as well as the main access to the port and the route for the three main highways that give access the city (Ponte Rio-Niterói, a bridge, Rodovia Washington Luiz and Rodovia Presidente Dutra, highways). The average daily flow of Avenida Brasil is 210 thousand vehicles, generating 770 calls for incidents on average per month.

The database used consists of 3,080 records that present the incident type, its location (georeferenced data), the time required and the vehicle involved. In addition, during the data treatment, it was identified that 282 records were incomplete. Thus, 2,798 records were used. Among them, 61.1% are broken-down vehicles, 16.7% are accidents and 9.2% are flat tires, see Fig. 3 (others represent incidents with a low frequency of events such as fires and being run over). In addition, 39.5% of the occurrences were serviced by motorcycles and 25.4% by light tows trucks, see Fig. 4.

4.1 Emissions and Exploratory Analysis

To reach the objectives of this study, it is necessary to calculate emissions for the real scenario, with the occurrence of incidents, and the emissions for a scenario without

Fig. 3 Occurrences by incident type

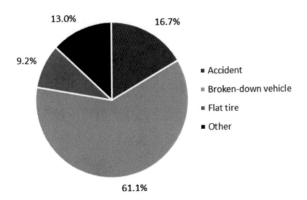

13.0%   16.7%

9.2%

- Accident
- Broken-down vehicle
- Flat tire
- Other

61.1%

Fig. 4 Occurrence by response vehicle

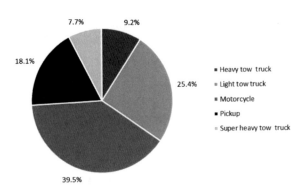

7.7%   9.2%

18.1%

25.4%

- Heavy tow truck
- Light tow truck
- Motorcycle
- Pickup
- Super heavy tow truck

39.5%

incidents (considering only the regular traffic flow). This strategy is used to determine the impact of incidents on the increase of CO_2 emissions, due to the occurrence of incidents.

According to Sect. 3, the input data used to estimate CO_2 emissions were: (i) traffic flow, obtained from the electronic surveillance system; (ii) the traffic composition, obtained from a manual counting; and (iii) the response time, calculated by the travel time of the tow truck between the section where it was located and the incident added to the time required to serve the vehicle.

As described in Sect. 3, emissions are calculated using the MEET model. For the scenario without incidents, it was considered the speed obtained using Google Maps for the time at which the incident occurred. For the scenario with the occurrence of incidents, a 40% reduction in traffic speed was imposed in the regular traffic condition obtained in Google Maps. This percentage was obtained considering a range of speed reduction due to the occurrence of incidents based on the literature. FHWA [17] showed that the reduction in speed is 75% and Haule et al. [24] showed a reduction of 25% in the same situation, so we decide to use 40%.

For the exploratory analysis, the data were organized considering three groups: (i) by incident type; (ii) by vehicle used to respond to the incident; and (iii) by the time of occurrence. It is worth mentioning that the duration of the incident is the time

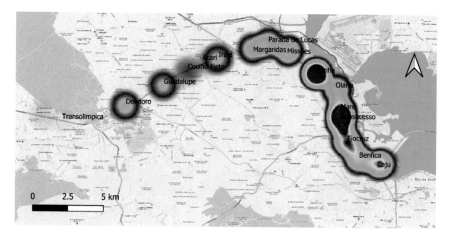

Fig. 5 Heat map of CO_2 emissions incidents in the scenario with incidents

elapsed between the moment that the vehicle obstructs the road and the moment when it is removed.

Thus, to start the data analysis, a Kernel Map was elaborated as shown in Fig. 5. Through this map, it is possible to identify the regions with higher CO_2 emissions due to incidents in the scenario with incidents. Studies to analyze incidents based on Kernel density estimation in a Geographic Information Systems environment are recurrent [8, 7, 30], as this analysis helps showing critical areas. In addition, its implementation is simple and easy to understand [8].

Based on the MEET results and using Kernel map, shown in Fig. 5, it is possible to identify the regions with the highest CO_2 emissions, in accordance with the incidents. It can be seen that the highest CO_2 emissions occur in regions close to the Penha, Bonsucesso, Fiocruz and Caju neighborhoods, areas that also have high volumes of vehicles involving trucks and buses. To carry out the analysis of emissions by incident type, these events were divided into four categories: (i) accidents; (ii) broken-down vehicles; (iii) flat tire; and (iv) others.

Therefore, the average, maximum and minimum rates, as well as the standard deviation, of CO_2 emissions by incident type are shown in Table 2. We can see that the incidents named as Others are the ones that most generate emissions because

Table 2 CO_2 emission (ton/km) per incident type

Incident type	Avg	Max	Min	Standard deviation
Accident	1.407406	11.14262	0.047608	1.189746
Broken-down vehicle	1.126868	11.4867	0.074229	0.83076
Flat tire	1.183577	6.06624	0.112714	0.842066
Other	1.780603	11.69225	0.122511	1.898567

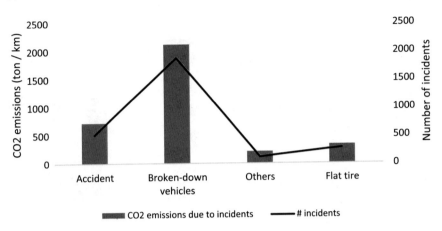

Fig. 6 Analysis per incident type

they have the longer average duration (32 min) compared to broken down vehicles (27 min), accidents (28 min) and flat tires (26 min). In relation to the maximum and minimum values, the highest emissions were observed in incidents that occurred during peak hours of the road, 8 am, 5 pm and 6 pm and the lowest emissions during the night, between 11 pm and 2 am.

We have computed the emissions of all incidents as shown in Fig. 6. We can see that broken-down vehicles are the main responsible for CO_2 emissions, due to their high frequency, even though these incidents cause the lowest average emissions (see Table 1). Incidents classified as others present a smaller emission when compared to the other types, despite being the ones that emit more in a unitary manner.

It was also considered pertinent to develop a Pareto Diagram, as shown in Fig. 7. This diagram shows the relationship between the incident type under analysis and the increase in CO_2 emissions due to its occurrence. On the left-vertical axis is presented, the difference between the emissions calculated in the region impacted by the incident and the emissions that would already exist due to the regular traffic flow, without the occurrence of the incident. Considering the sum of emissions resulting from the occurrence of the incidents, it is observed that broken-down vehicles are responsible for 63.3% of the increase in CO_2 emissions. When we add broken-down vehicles and accidents, it represents 84.2% of the emissions.

In order to represent what was exposed previously, Fig. 8 presents Kernel maps for each incident type, showing that broken-down vehicles are the incidents responsible for the largest contribution of emissions, followed by accidents and flat tires. Furthermore, in the four maps presented, a higher concentration of CO_2 emissions was identified between Penha and Caju neighborhoods, as already noted in Fig. 5.

To carry out the analysis of emissions by vehicle type used to respond the incidents, these vehicles were divided into five categories: (i) motorcycle; (ii) pickup; (iii) light tow trucks (iv) heavy tow trucks; and (v) super heavy tow trucks. It is worth mentioning that motorcycles serve any vehicle type, since they do the detection and

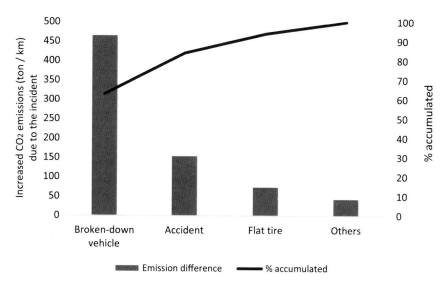

Fig. 7 Pareto Analysis considering the increasing in CO_2 related to the scenario without incidents

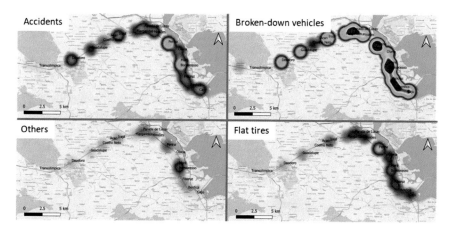

Fig. 8 Heat Map of CO_2 emissions by incident type

beaconing, and that the pickups also help in the detection, but only remove passenger cars from the road, as well as light tow trucks. In addition, heavy tow trucks serve buses and smaller trucks, and the super heavy tow trucks serve just big trucks. Table 3 shows the CO_2 emission per vehicle type used.

With these results, we can see that the assistances provided by heavy tow trucks, followed by those assisted by super heavy tow trucks, are the ones that have the highest average CO_2 emissions. Considering the maximum values, when analyzing the database, we can note that the highest emissions in four cases were observed in

Table 3 CO_2 emission (ton/km) per vehicle type used on responses

Vehicle type	Avg	Max	Min	Standard deviation
Heavy tow	1.377804	6.787573	0.331819	0.903166
Light tow	1.230352	8.069689	0.047608	0.892012
Motorcycle	1.263497	11.69225	0.171284	1.086791
Pickup	0.956686	10.74192	0.096552	0.881699
Super heavy tow	1.323073	7.747977	0.057445	1.047302

incidents that occurred during the peak hours of the road, at 8 am and at 5 pm. This was not observed in the maximum value observed in incident that used the super heavy tow truck on response, in this case, the highest emission is from an accident with more than 4 h to clear. It is important to mentioning that the minimum emission values were observed in incidents that occurred at dawn, when the traffic flow is reduced.

However, when we consider all responses provided by the vehicles, Fig. 9 shows the total CO_2 emissions by vehicle type. The data indicate that motorcycles are the vehicles that most attend events (39.5%), followed by light tows trucks (25.4%). Therefore, the response performed by these two vehicles presents the higher CO_2 emission.

In a complementary way, the time of the incident directly influences the CO_2 emissions, since the vehicular flow is higher during peak hours and, consequently, the congestion queue formed generates larger negative impacts. Thus, Fig. 10 shows the CO_2 emissions related to incidents considering the time of the day. The results indicate that emissions are concentrated between 6 am and 7 pm, with the highest peaks at 6 am and at 6 pm. In addition, it is noted that, as expected, the lowest emissions are concentrated at dawn. Without incidents, the same period, between

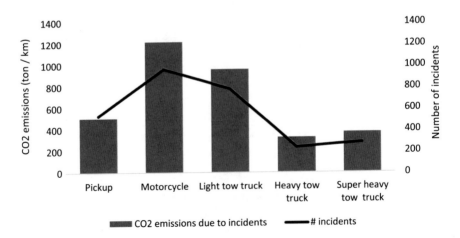

Fig. 9 Analysis by vehicle used on responses

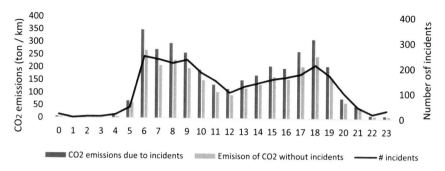

Fig. 10 Analysis per hour of incident

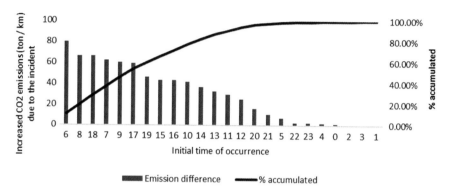

Fig. 11 Pareto Analysis considering the time of the occurrence

6 am and 7 pm, also concentrates more emission. Therefore, Fig. 11 shows that incidents that occur between 6 am and 10 am and between 2 pm and 7 pm are responsible for 82.4% of the increase in CO_2 emissions in Avenida Brasil.

We have compared the scenario with occurrence of incidents, with another one that does not consider the occurrence of incidents on the road (considering only the regular traffic flow), as summarized in Fig. 12. It is possible to note that there is a reduction in CO_2 emissions when considering the scenario without incidents as expected. This

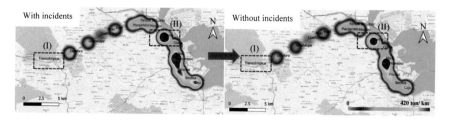

Fig. 12 Heat Map based on CO_2 emissions for the scenarios with and without occurrence of incidents

reduction represents a 22% drop in CO_2 emissions. In addition, looking at the areas represented on the map, it is identified that: (i) the greatest reduction in emissions was observed in Transolimpica region; and (ii) a greater reduction in the sum of all incidents was identified in Penha region, since the highest number of occurrences recorded was in this region.

5 Discussion of Policies and Actions to Minimize Incidents

The establishment of public policies and mitigation actions that really control the concentration of air pollutants in urban areas has been a major challenge for decision-makers [32]. It happens because the process to develop efficient strategies, which reduces pollutant emissions in the field of transportation, must consider the participation of all actors involved in the operation of the sector, such as public agencies, transport planners, drivers, pedestrians, public transport concessionaires and logistics companies. However, as shown in the results discussed above, there is a need to implement policies and actions to reduce the problem related to the increase in CO_2 emissions from road incidents.

According to Haque et al. [23], to increase traffic safety and thus reduce road incidents such as traffic accidents, both drivers and pedestrians need to understand their social responsibility and public institutions need to legislate to reach a higher level of security. Therefore, Welle et al. [40] indicate actions that can be implemented to create a safe mobility system on the roads presented as follow: (i) build compact and connected cities; (ii) design more intelligent routes; (iii) offer a variety of secure mobility options; (iv) keep speeds at safety levels; (v) enforce existing laws and regulations; (vi) provide a better education for drivers and urban planners; (vii) require universal vehicle safety standards; and (viii) accelerate emergency response.

Incident reduction initiatives may include the use of Intelligent Transportation Systems (ITS), which allows, for example, a better coordination between those responsible for the incident management and the use of mobile assistance services to drivers on the roads. According to Chen et al. [12], energy efficiency and emissions reduction have become essential arguments for investments in ITS. In addition, the authors indicate that, for cities in developing countries, the focus of these investments should be on basic infrastructure, including a coherent road network and basic traffic management measures. Therefore, investment in ITS is an important strategy in order to reduce the negative externalities generated by the incidents. We can focus on automatic incident detection systems, which have algorithms that detect the occurrence of incidents by processing data related to traffic parameters, such as traffic flow, average speed and occupancy rate [11].

Another measure that can be adopted, this time to reduce the total time of the incident, is the repositioning of the service vehicles, as studied by Geroliminis et al. [20], Ozbay et al. [34], Zhu et al. [43], Adler et al. [2] and Baltar et al. [5]. According to Yuan and Cheu [41], the losses and the influence of an incident are proportional to its duration, therefore, a quick detection and an effective service can considerably reduce the negative impacts of the incidents. Zhu et al. [43], for example, developed a model that seeks to reduce the total service time by allocating patrols, considering the possibility of occurring up to two incidents at the same time, with total time corresponding to the sum of the first and second service times. In addition, this paper defined a method for delimiting the service areas of each patrol. Complementarily, Adler et al. [2] studied the location of patrols that attend incidents and inspections of the road. The authors proposed a mathematical model, which considers that the patrols must be close to the accidents and close to places where many traffic violations occur. For the case of Avenida Brasil, Baltar et al. [5] explored the trade-off between maximizing the handling of traffic incidents and minimizing CO_2 emissions to locate tow vehicles.

It is also worth mentioning that it is possible to focus on vehicle technologies such as autonomous driving to reduce incidents. Autonomous vehicles (AVs) have been considered the sustainable future for providing greater road safety, efficient traffic flow and lower fuel consumption and, consequently, lower CO_2 emissions, improving urban mobility and, therefore, general well-being [9, 31]. AVs have the potential to significantly mitigate many of the errors that drivers routinely make [3, 16, 35] and then improve road safety consequently. The reason is that they have better perception (for example, absence of blind spots), better decision-making (for example, more accurate planning of complex steering maneuvers, such as parallel parking) and better execution (for example, faster and more precise control of steering, brakes and acceleration) [28].

In relation specifically to the results of this research, there is an expressive occurrence of broken-down vehicles that makes the incident types the main responsible for the increase in CO_2 emissions. Therefore, public policies that increase the inspection of vehicle conditions and more rigorous annual inspections need to be encouraged. In addition, the need for better robustness in operations for detecting and responding to incidents is identified, mainly between 6 am and 10 am and between 2 pm and 7 pm, which are in the morning and afternoon peak periods, respectively. It is noteworthy that these periods together are responsible for 82.4% of the increase in CO_2 emissions in the region under analysis when compared to the scenario without the occurrence of incidents. In addition, heavy and super heavy tow trucks handle incidents that generate higher emissions, on average. The reason is that they are responsible for serving trucks and big trucks, responsible for freight transportation. Therefore, the adoption of a freight vehicle restriction policy during peak periods of the road may be interesting in order to reduce incidents involving this vehicle type.

6 Conclusions

Daily traffic congestions affect the quality of life of city-dwellers, due to travel time delays, excessive fossil fuel consumption and increased environmental pollution. Congestion can also be intensified through traffic incidents that generate frequent and irregular road interruptions. Thus, studies that aim to minimize the impacts of incidents need to be done. Therefore, this chapter serves this purpose by analyzing, through a case study of Avenida Brasil (the main expressway in Rio de Janeiro city), the impacts on CO_2 emissions as well as proposing actions to minimize these emissions. The methodology used the Methodologies for Estimating air pollutant Emissions from Transport (MEET), model proposed by Hickman et al. [25], to calculate the emission of CO_2, which is based on vehicle type and on speed.

The results indicate that the incidents contribute to a 22% increase in CO_2 emissions, that the broken-down vehicles are the incidents with the greatest impact on these emissions, due to its high frequency, despite being the incidents that on average cause less emission on the road, and that the peak hours in the morning (between 6 am and 10 am) and in the afternoon (between 2 pm and 7 pm) are responsible for 82.4% of the increase in CO_2 emissions, related to the incidents. It is also worth mentioning that, with the analysis of the maps generated through the Kernel density estimator, it is possible to verify the sections with the highest occurrences of incidents, as well as CO_2 emissions.

Having exposed the results, this chapter also indicates proposals for improving the framework presented through public policies and specific actions that can help to minimize the impacts of incidents, in a more general context which include: (i) increase traffic safety by creating a safe road mobility system; (ii) the use of Intelligent Transportation Systems (ITS), which allows, for example, a better coordination between those responsible for incident management and the use of mobile services to assist drivers on the roads; (iii) repositioning of service vehicles that allows rapid detection and effective service; and (iv) focus on vehicle technologies such as autonomous driving that have the potential to significantly mitigate many of the mistakes that drivers routinely make.

In addition, when a more accurate diagnosis of the situation is made, through analysis of problematic regions, as done specifically for Avenida Brasil, efforts should be made to address the specific problems found. For example, it was noted for Avenida Brasil an expressive contribution of broken-down vehicles, increasing the CO_2 emissions, thus policies such as rigorous annual vehicle inspections must be encouraged to minimize the problem.

This study is an important tool to assist decision-makers, responsible for incident management, regarding the concentration of efforts to improve air quality and, consequently, the life of the urban population. With a more effective incident management and the reduction of CO_2 emissions, the main contributor to the increase in greenhouse gases (GHG) emissions, it is possible to: (i) ensure a healthy life and promote well-being for all, at all ages, due to the minimization of the social impact caused by the incidents such as increased travel time, stress, pollutant inhalation and others; (ii) making cities and human settlements inclusive, safe, resilient and sustainable, due to the faster response to incidents, which reduces secondary accidents, thereby increasing road safety; (iii) ensuring sustainable production and consumption patterns, due to the reduction of congestion that impacts on fuel consumption; and (iv) take urgent measures to mitigation of climate change and its impacts. It is noteworthy that all these actions are in line with the goals and targets of the 2030 Agenda for Sustainable Development [39].

In view of the needs for further studies on the subject, we recommend to apply the methodology throughout the Rio de Janeiro city, not only in an express road as carried out in this chapter, as well as to apply it to other areas in Brazil and in developing countries that suffer from the loss of quality of life of the urban population due to the high CO_2 emissions.

Acknowledgements This work was partially supported by the National Council for Scientific and Technological Development (CNPq), under grant #307835/2017-0. This work was supported by Carlos Chagas Filho Foundation for Research Support of the State of Rio de Janeiro, under grants #233926. This study was also financed in part by the Coordenação de Aperfeiçoamento de Pessoal de Nível Superior—Brasil (CAPES)—Finance Code 001. We would like to thank CET-Rio for providing the database for this research.

References

1. Abou-Senna H, Radwan E (2013) VISSIM/MOVES integration to investigate the effect of major key parameters on CO_2 emissions. Transp Res Part D: Trans Environ 21:39–46. https://doi.org/10.1016/j.trd.2013.02.003
2. Adler N, Hakkert AS, Kornbluth J, Raviv T, Sher M (2013) Location-allocation models for traffic police patrol vehicles on an interurban network. Ann Oper Res 221(1):9–31. https://doi.org/10.1007/s10479-012-1275-2
3. Anderson JM, Kalra N, Stanley KD, Sorensen P, Samaras C, Oluwatola OA (2016) Autonomous vehicle technology: a guide for policymakers. RAND Corporation, RR-443-2-RC, Santa Monica, Calif. Available at: http://www.rand.org/pubs/research_reports/RR443-2.html
4. Bai X, Zhou Z, Chin K-S, Huang B (2017) Evaluating lane reservation problems by carbon emission approach. Trans Res Part D: Trans Environ 53:178–192. https://doi.org/10.1016/j.trd.2017.04.002
5. Baltar M, Abreu V, Ribeiro G, Bahiense L (2020) Multi-objective model for the problem of locating tows for incident servicing on expressways. TOP. https://doi.org/10.1007/s11750-020-00567-w
6. Barth M, Boriboonsomsin K (2008) Real-World carbon dioxide impacts of traffic congestion. Trans Res Record: J Trans Res Board 2058(1):163–171. https://doi.org/10.3141/2058-20
7. Bíl M, Andrásik R, Janoska Z (2013) Identification of hazardous road locations of traffic accidents by means of kernel density estimation and cluster significance evaluation. Accid Anal Prev 55:265–273. https://doi.org/10.1016/j.aap.2013.03.003
8. Blazquez C, Celis M (2011) A spatial and temporal analysis of child pedestrian crashes in Santiago, Chile. Accident Anal Prevent 50:304–311. https://doi.org/10.1016/j.aap.2012.05.001
9. Burns LD (2013) A vision of our transport future. Nature 497:181–182. https://doi.org/10.1038/497181a
10. Chen K, Yu L (2007) Microscopic traffic-emission simulation and case study for evaluation of traffic control strategies. J Transp Syst Eng Inf Technol 7(1):93–99. https://doi.org/10.1016/s1570-6672(07)60011-7
11. Chen L, Cao Y, Ji R (2010) Automatic incident detection algorithm based on support vector machine. IEEE Sixth International conference on natural computation, 864–866. https://doi.org/10.1109/icnc.2010.5583920
12. Chen Y, Gomez A, Frame G (2017) Achieving energy savings by intelligent transportation systems investments in the context of smart cities. Transp Res Part D: Trans Environ 54:381–396. https://doi.org/10.1016/j.trd.2017.06.008

13. Chung Y, Cho H, Choi K (2013) Impacts of freeway accidents on CO_2 emissions: a case study for Orange County, California, US. Trans Res Part D: Trans Environ 24:120–126. https://doi.org/10.1016/j.trd.2013.06.005

14. DECC—Department of Energy & Climate Change (2014) 2012 UK Greenhouse Gas Emissions. Final Figures. Department of Energy & Climate Change, London, UK

15. De Palma A, Lindsey R (2011) Traffic congestion pricing methodologies and technologies. Trans Res Part C: Emerg Technol 19(6):1377–1399. https://doi.org/10.1016/j.trc.2011.02.010

16. Fagnant DJ, Kockelman K (2015) Preparing a nation for autonomous vehicles: opportunities, barriers and policy recommendations. Transp Res Part A: Policy Pract 77:167–181. https://doi.org/10.1016/j.tra.2015.04.003

17. FHWA—Federal Highway Administration. (2004). Incident characteristics and impact on freeway traffic. Report n° FHWA/TX-05/0-4745-1

18. Figliozzi MA (2011) The impacts of congestion on time-definitive urban freight distribution networks CO_2 emission levels: results from a case study in Portland, Oregon. Transp Res Part C: Emerg Technol 19(5):766–778. https://doi.org/10.1016/j.trc.2010.11.002

19. Fogliatti MC, Mattos NMC (2006) Teoria de filas (Queuing theory). Editora Interciência (Publisher Interciência), Rio de Janeiro City

20. Geroliminis N, Karlaftis MG, Skabardonis A (2009) A spatial queuing model for the emergency vehicle districting and location problem. Transp Res Part B: Methodol 43(7):798–811. https://doi.org/10.1016/j.trb.2009.01.006

21. Guerrieri M, Mauro R (2016) Capacity and safety analysis of hard-shoulder running (HRS): a motorway case study. Transp Res Part A: Policy Pract 92:162–183. https://doi.org/10.1016/j.tra.2016.08.003

22. Grote M, Williams I, Preston J, Kemp S (2016) Including congestion effects in urban road traffic CO_2 emissions modelling: Do Local Government Authorities have the right options? Transp Res Part D: Transp Environ 43:95–106. https://doi.org/10.1016/j.trd.2015.12.010

23. Haque M, Chin H, Debnath A (2013) Sustainable, safe, smart—three key elements of Singapore's evolving transport policies. Transp Policy 27:20–31. https://doi.org/10.1016/j.tranpol.2012.11.017

24. Haule HJ, Sando T, Lentz R, Chuan C-H, Alluri P (2019) Evaluating the impact and clearance duration of freeway incidents. Int J Transp Sci Technol. https://doi.org/10.1016/j.ijtst.2018.06.005

25. Hickman J, Hassel D, Joumard R, Samaras Z, Sorenson S (1999) Methodology for calculating transport emissions and energy consumption. Int J Transp Sci Technol 8:13–24

26. IPCC—Intergovernmental Panel on Climate Change. (2014). Climate Change 2014: Mitigation of Climate Change. Contribution of Working Group III to the Fifth Assessment Report of the Intergovernmental Panel on Climate Change, Cambridge University Press, Cambridge, United Kingdom and New York, NY, USA

27. Jabali O, Van Woensel T, de Kok AG (2012) Analysis of travel times and CO_2 emissions in time-dependent vehicle routing. Product Oper Manag 21(6):1060–1074. https://doi.org/10.1111/j.1937-5956.2012.01338.x

28. Kalra N, Paddock SM (2016) Driving to safety: how many miles of driving would it take to demonstrate autonomous vehicle reliability? Transp Res Part A: Policy Pract 94:182–193. https://doi.org/10.1016/j.tra.2016.09.010

29. King D, Inderwildi O, Carey C, Santos G, Yan X, Behrendt H, Holdway A, Owen N, Shirvani T, Teytelboym A (2010) Future of mobility roadmap—ways to reduce emissions while keeping mobile. Smith School of Enterprise and the Environment, Oxford, UK

30. Kingham S, Sabel C, Bartie P (2011) The impact of the 'school run' on road traffic accidents: A spatio-temporal analysis. J Trans Geograph 19(4):705–711. https://doi.org/10.1016/j.jtrangeo.2010.08.011

31. Le Vine S, Zolfaghari A, Polak J (2015) Autonomous cars: the tension between occupant experience and intersection capacity. Transp Res Part C: Emerg Technol 52:1–14. https://doi.org/10.1016/j.trc.2015.01.002

32. Maes A, Hoinaski L, Meirelles T, Carlson R (2019) A methodology for high resolution vehicular emissions inventories in metropolitan areas: Evaluating the effect of automotive technologies improvement. Transp Res Part D: Transp Environ 77:303–319. https://doi.org/10.1016/j.trd.2019.10.007

33. Muthu SS (2019) Carbon footprints: case studies from the energy and transport sectors. Springer, Environmental Footprints and Eco-design of Products and Processes

34. Ozbay K, Iyigun C, Baykal-Gursoy M, Xiao W (2012) Probabilistic programming models for traffic incident management operations planning. Ann Oper Res 203(1):389–406. https://doi.org/10.1007/s10479-012-1174-6

35. Pereira A, Anany H, Pribyl O, Prikryl J (2017) Automated vehicles in smart urban environment: A review. IEEE smart city symposium prague (SCSP). https://doi.org/10.1109/scsp.2017.7973864

36. Smit R, Brown AL, Chan YC (2008) Do air pollution emissions and fuel consumption models for roadways include the effects of congestion in the roadway traffic flow? Environ Model Softw 23(10–11):1262–1270. https://doi.org/10.1016/j.envsoft.2008.03.001

37. Sookun A, Boojhawon R, Rughooputh SDDV (2014) Assessing greenhouse gas and related air pollutant emissions from road traffic counts: a case study for Mauritius. Transp Res Part D: Trans Environ 32:35–47. https://doi.org/10.1016/j.trd.2014.06.005

38. Tiwary A, Chatterton T, Namdeo A (2013) Co-managing carbon and air quality: pros and cons of local sustainability initiatives. J Environ Plann Manag 57(8):1266–1283. https://doi.org/10.1080/09640568.2013.802677

39. United Nations (2015) Transforming our world: the 2030 agenda for sustainable development—A/RES/70/1

40. Welle B, Sharpin AB, Adriazola-Steil C, Job S, Shotten M, Bose D, Bhatt A, Alveano S, Obelheiro MR, Imamoglu CT (2018) Sustentável e Seguro. Visão e Diretrizes para Zerar as Mortes no Trânsito (Sustainable and Safe. Vision and Guidelines for Zero Traffic Deaths). WRI Ross Center for Sustainable Cities e Global Road Safety Facility. https://wribrasil.org.br/sites/default/files/Sustentavel_Seguro.pdf

41. Yuan F, Cheu RL (2003) Incident detection using support vector machines. Transp Res Part C: Emerg Technol 11(3–4):309–328. https://doi.org/10.1016/s0968-090x(03)00020-2

42. Zhang K, Batterman S, Dion F (2011) Vehicle emissions in congestion: Comparison of work zone, rush hour and free-flow conditions. Atmos Environ 45(11):1929–1939. https://doi.org/10.1016/j.atmosenv.2011.01.030

43. Zhu S, Kim W, Chang G-L, Rochon S (2014) Design and evaluation of operational strategies for deploying emergency response teams: dispatching or patrolling. J Transp Eng 140(6):04014021. https://doi.org/10.1061/(asce)te.1943-5436.0000670

Carbon Footprint Estimation for Academic Building in India

Venu Shree, Himanshu Nautiyal, and Varun Goel

Abstract Carbon footprint is the process in which the effect of estimation of carbon emission in a particular system/process is evaluated from its origin (cradle) to its death (grave). In industry, carbon emissions released due to various activities like using materials, transportation, utility equipment etc. Carbon footprint helps us in the identification of carbon-intensive processes and also helps us in the comparison among different products/processes. Nowadays, due to rapid development of cities, massive amount of emissions are happening in all the sectors. In the present chapter, emissions associated with an academic building are estimated using LCA technique in GaBi education software. After assessing its carbon footprint, some remedial measures were also introduced to reduce its carbon footprint and make the system more sustainable. The study also helps us to know the effects of various parameters of a building that affects the emission level and try to modify it in accordance with having fewer emissions.

1 Introduction

The world is acutely adapting various sustainable measures to mitigate the problem of climate change. Even if climate change is the part of natural earth processes, there is huge augmentation of various domestic and industrial activities of human beings, which accelerate the changes in earth's climate. It is being observed throughout the

V. Shree
Department of Architecture, National Institute of Technology Hamirpur, Hamirpur, HP, India
e-mail: venushree80@gmail.com

H. Nautiyal (✉)
Department of Mechanical Engineering, THDC Institute of Hydropower Engineering and Technology, Tehri, Uttarakhand, India
e-mail: h2nautiyal@gmail.com

V. Goel
Department of Mechanical Engineering, National Institute of Technology Hamirpur, Hamirpur, HP, India
e-mail: varun7go@gmail.com

© The Author(s), under exclusive license to Springer Nature Singapore Pte Ltd. 2021
S. S. Muthu (ed.), *LCA Based Carbon Footprint Assessment*, Environmental Footprints and Eco-design of Products and Processes, https://doi.org/10.1007/978-981-33-4373-3_3

world that Greenhouse Gas (GHG) emissions produced by these human activities are the main cause of climate change. High amount of GHG emissions are being released into the atmosphere, which promotes the climate change issues in the form of several environmental impediments like global warming, ozone layer depletion, acidification etc. Apart from these environmental problems, these GHG emissions are also responsible for deteriorating fresh air quality and health of human beings. In order to control these environmental issues, new measures must be incorporated to reduce the production of GHG emissions into the atmosphere. Control of GHG emissions is also important to achieve emissions reduction targets under Clean Development Mechanism (CDM) introduced in Kyoto Protocol [13, 16]. Carbon dioxide (CO_2), methane, nitrous oxide, ozone, fluorinated gasses etc. are the GHG gases present in earth's atmosphere. CO_2 is the main constituent of GHG and released through various human activities involving combustion of fossil fuels. In addition to these, human activities are also responsible for the reduction in carbon sequestration capacity of earth. As the proportion of CO_2 in the earth's atmosphere should be stable but it is continuously increasing speedily. The amount of CO_2 emissions in earth's atmosphere is increased with the augmentation of population, urbanization and their energy needs [11]. The extensive use of conventional fossil fuels throughout the world releases high amount of GHG emissions into the atmosphere. Although almost all nations of the world are trying to increase renewable energy-based power generation to minimize the dependency on conventional energy sources and to promote sustainable energy sources. Consequently, it becomes quite essential to minimize carbon emissions in all these important sectors. Building sector is one of the main sectors, which are having contribution in release of high amount of GHG emissions in the atmosphere. Large-scale building construction and other relative activities, transportation etc. are having a significant role in high discharge of GHG emissions. Figure 1 shows global emissions associated with building sector in past years (Fig. 2).

Building is one of the vital amenities of human beings. Buildings are constructed with the objectives to provide shelter and appropriate indoor environment. Along with the progress of human civilization, a huge advancement has been seen in building construction. In the present era, various types of building structures with different shapes and sizes are being constructed throughout the world using advance sophisticated techniques of construction, variety of building materials and human skills. Buildings can be assorted as residential and commercial depending upon their use and operation. Residential buildings include small and big houses whereas offices, shops, schools, colleges etc. are associated with the commercial buildings [15]. The initial phase of a building begins with the construction processes, which include significant planning, design, investment and other legal considerations [26]. After construction and commissioning of the building, a lot of energy is utilized in operational phase in electrical and electronic systems and appliances, heating, ventilation and air conditioning (HVAC), lightings etc. Finally, the demolition of building is carried out to tear down the building after completing its life. Maintenance and repair is another important aspect and must be included in operational phase of a building life cycle [29]. A lot of energy is consumed by buildings throughout their entire life cycle [4] and generate high amount of GHG emissions. In order to analyze energy consumption in

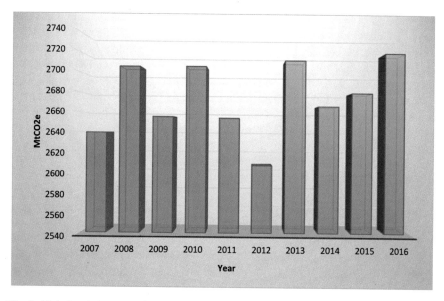

Fig. 1 Global emissions associated with building sector (www.wri.org) [33]

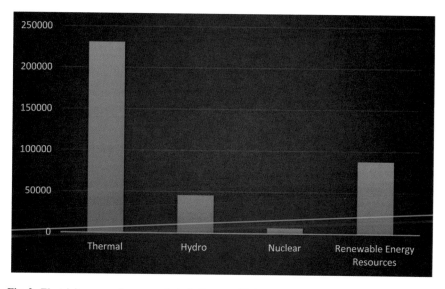

Fig. 2 Electricity generation scenario in India as on 30 June 2020 (www.cea.nic.in) [3]

buildings and associated emissions, the life cycle of a building can be divided into three primary phases: construction, operation and maintenance; and finally demolition. In all these phases, buildings consumed high amount of energy and produce GHG emissions, which are continuously increasing at massive rate [5, 24]. The consumption of energy in buildings during construction, operation and demolition stages is called direct energy. The production of materials associated with different construction activities and building installations involves high energy consumption [21]. Therefore, buildings have significant contribution in global GHG emissions in atmosphere and due to this, it becomes essential to study and investigate all those factors associated with high emerge consumption and release of GHG emissions in buildings. Various studies also advocate to explore and study a building along with its various elements of before construction process in order to achieve the goals of sustainable development [25]. Consequently, building sector is now trying to develop and adopt new design and construction techniques, which makes the buildings energy efficient and reduces GHG emissions during all phases of building's life cycle. In addition to this, recycling and reuse of waste building materials also help in reducing the need for fresh materials in new building construction and thus help in minimizing primary energy requirement and emissions [14].

In order to achieve the goal of reducing GHG emissions associated with buildings, it is important to estimate and quantify them. The use of "carbon footprints" is quite useful and popular terminology to quantify GHG emissions and may be defined as the quantity of CO_2 released by a product and its associated processes over all phases of its life cycle. It is derived from the term "ecological footprint," which was introduced by Wackernagel and Rees [31]. Carbon footprints can also be defined as the amount of CO_2 emissions directly and indirectly associated with an activity or a product over its entire life cycle; and expressed as grams of CO_{2eq} per kilowatt hour (gCO_{2eq}/kWh) and is also used for accounting global warming effects of other GHGs [32].

Further, effective methodologies are required to estimate carbon footprints associated with an activity or a product. Life Cycle Assessment (LCA) is one of the popular methods to analyze environmental impacts associated with an activity, product or process throughout its life cycle [8, 12]. It is also considered as versatile and effective tool for the estimation and reduction of energy consumption and associated emissions of the building sector [2, 6]. Advent of software further improved the speed and reliability of calculation and results of LCA, sustainability assessment and comparison of environmental impacts associated with various products. Therefore, environmental impacts of different types of buildings can be studied effectively using LCA methodology [30]. In the present chapter, an educational building is considered to be studied and its associated carbon footprints are estimated using LCA. The educational buildings involve so many energy intensive activities and it is important to study their environmental performance for sustainable development in future. In fact, various environmental management system and policies are being introduced to mitigate environmental impacts associated with educational infrastructures [6]. The chapter begins with the introduction part with discussion related to emissions associated with building sector and need to control them. Section 2 discusses about the LCA method used to study environmental impacts of a building. In Sect. 3, methodology used to carry out to estimate emission associated with a building selected for case study using LCA is discussed. Section 4 discusses the detailed case study and estimation of emissions which is then followed by discussion and calculations discussed in Sects. 5 and 6 respectively.

2 Life Cycle Assessment

LCA is a methodology that can be used to assess utilized materials, various energy flows and environmental impacts associated with the products. It helps in quantitative assessment of all materials and processes in a systematic way. It is also effective in study of various features and aspects of development of life of a product (ISO, 1997)[9]. Another important feature of LCA methodology is that it is a multidisciplinary approach in the way that natural environmental impacts and human relations to these impacts can be modeled. LCA can also be classified into Process LCA, Input–Output LCA and Hybrid LCA. The methodological framework of LCA has four main parts. These are goal and scope definition, life cycle inventory analysis, life cycle impact assessment and life cycle interpretation [23, 27, 29]. In the first stage of goal and scope definition, functional unit, application, system boundaries etc. are chosen. Collection of data, building of system model is associated with life cycle inventory analysis phase. The next phase, i.e., life cycle impact assessment involves classification of different inventory parameters in accordance with the type of environmental impacts like GHG emissions, acidification, water footprints, etc. [17]. A characterization factor is also introduced to express inventory results into potential environmental indicators. The last phase is life cycle interpretation in which results obtained from life cycle inventory analysis and life cycle impact assessment are interpreted and further used in recognition of important issues and assessment of results [23, 28].

In order to study environmental performance of buildings, LCA is a quite effective methodology and it helps in performing detailed analysis of various environmental impacts associated with a building throughout its complete life cycle [10]. It is also helpful in quantification and comparison of different buildings in context with their environmental performance and energy use. It serves as an effective technique to find and study various factors associated with a building's life cycle where improvements are required to reduce the environmental impacts. Now days, performing LCA using software further makes the data collection process simpler and less time consuming, which improves the quality of results. Software like SimaPro, GaBi, Open LCA, Umberto etc. are quite popular to perform LCA on various products and process. Carbon footprints can be effectively estimated using LCA method and this estimation can be further improved and performed in shorter time using software. As already discussed, LCA requires end to end analysis of a product, which is carried out by considering all raw materials used and their transportation, operation and disposal. Estimation of carbon footprints of a product or a service is an important outcome of LCA, which gives a broad picture of GHG emissions associated with the product or service.

3 Methodology

3.1 Case Study

In the present study, carbon footprints are estimated of an educational building at THDC Institute of Hydropower Engineering & Technology, Tehri, Uttarakhand, India. As per the definition, the complete life cycle of a building is classified into three different phases, viz. construction phase, operation and maintenance phase and demolition phase. In the present study, construction and operation maintenance phases of the building are taken into account for carbon footprint calculation. In fact, the energy required to demolish the building is quite less and may be considered about 1–2% of the life cycle energy of the building [19] but the demolition of building is not quite common so it is excluded from the present study. The operational time of the building is considered to be 50 years. In addition to this, any modifications and reconstruction works are excluded throughout the life cycle of the building. As variety of raw materials are used in construction of a building and these materials are produced by various process and methods with sufficient amount of energy consumption. So, every construction material is associated with primary energy consumption and release of GHG emissions. As far as the materials used in the construction of the selected building are concerned, problem of unavailability of data is faced during inventory analysis. Various types of materials are used in construction of a building but it was difficult to find accurate amount of these construction materials used in buildings. Therefore, the main construction materials viz. Cement, sand and bricks are considered in the study. Due to unavailability of on-site construction data, transportation activities are excluded from the present study.

In operational phase, electricity consumption is estimated using previous electricity bills. An assumption of same annual operational energy of the building is considered throughout its entire life cycle, however, it may be changed due to variation in external conditions. The electricity is dedicatedly used for lighting, running of PCs and other small lab equipment. The building is not having any heating, ventilation and air conditioning (HVAC) system and therefore not considered in the present analysis. The electricity scenario in India is presented in Fig. 2. Moreover, variation in primary energy consumption and emissions due variation in climatic conditions are not included. In addition to the operation phase of a building, maintenance and other repair works are important parts of building life cycle and these activities also consume a lot of energy and associated with release of GHG emissions. As the selected building is commissioned in year 2011 and there are no maintenance records found for the building and excluded from the study.

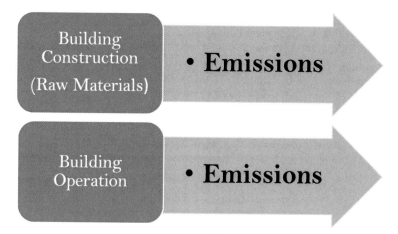

Fig. 3 System boundaries

3.2 Functional Unit

In order to carry out LCA of the selected building, the important step is to select the functional unit which may be the floor area, rooms etc. Along with selection of functional unit, relation of inputs and outputs associated with a system should be specified clearly. In the present analysis "m³ volume" of the building is considered and the complete service life of the building is considered to be 50 years. Moreover, the annual operational energy consumption throughout its complete cycle is assumed to be uniform. The emissions associated with the building are estimated considering an assumption that no reconstruction, extension and other minor civil works will be carried out during its complete life cycle. Maintenance, reconstruction and other modifications activities in building increase resources and energy consumption in building. However, they are quite less as compared with resources and energy consumption associated with construction and operational phase of the building. So, the main emphasis in the present case study is given on construction and operational phase of the building only (Fig 3). The analysis to estimate emissions is done using GaBi Education software (Version 9.2.1).

3.3 Limitations

The present study comprises certain limitations as many assumptions and simplifications are considered due to unavailability of ample data. Many construction materials are not considered in the analysis as on site construction data was not available for the selected building and therefore only important construction materials viz. cement, bricks and sand are considered in the study. Various human factors like working environment, odor etc. are excluded too. Moreover, assets like furniture, cabinets and other small accessories are not included in the study but electricity consumption in lighting, PCs, fans and other lab equipment is considered in the operational phase of the building. The study is set about with collection and quantification of construction materials used in the building. The data and details related to materials using construction of the building are obtained by using visual inspection and building's drawing and plans shown in Figs. 4, 5 and 6. The electrical energy consumed in the operational phase of the building is found by electricity bills of previous bills. As the selected building is an academic building and is nonresidential type so the electricity loads are assumed to be constant with respect to a specific schedule. These obtained data are used in GaBi software to estimate emissions associated with the building.

4 Case Study

The building selected for the case study is an educational building of THDC Institute of Hydropower Engineering and Technology, which is located at hilly region of Tehri, Uttarakhand India. The coordinates of the location are 30.38 latitude and 78.46 longitude. The building is located near to Tehri Hydropower reservoir, which is one of the largest hydropower projects in the world with generation capacity of 2400 MW. The building was commissioned in 2011 and dedicatedly used for academic purpose. The building comes under the Department of Applied Sciences and having labs, classrooms, computer labs, faculty cabins, etc. Figure 7 represents the actual picture of the building. The building is having two floors and floor area of the building is about 1112 m^2. The objective of the present study is to estimate carbon footprints of the selected building using LCA technique with the help of software. LCA studies are able to consider various environmental impacts associated with the building from acquisition of raw materials to its disposal as per ISO 14040 2006. In the present study, process-based LCA technique is used. The analysis is started with a selection of building and then the resources associated directly as well as indirectly to produce it are analyzed, which is then followed by estimation of carbon footprints.

The location of the building is in the hilly terrains of Tehri and variations in climatic conditions are experienced throughout the year. But as it has been discussed in previous section that the effect of these climatic conditions is not considered in the study. The building comprises physics lab, chemistry labs, faculty cabins, administrative offices in ground floor and two classrooms and computer labs in the first floor. Electrical energy is dedicatedly consumed in lighting, fans, computers, heaters, sanitation equipment, office equipment like projectors, photocopiers and other lab equipment. In order to calculate annual electrical consumption of the building, it is quite obvious to depend on the previous electricity bills. The annual electricity consumption is calculated using the electrical ratings found in previous electricity bills and further extended to total electricity consumption throughout its life cycle i.e. 50 years. From this method, the annual electricity consumption of the selected building is about 4945.70 kWh$_e$/year.

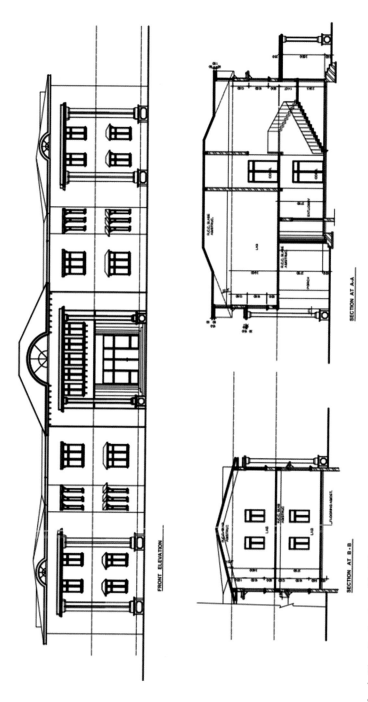

Fig. 4 Elevation/section of building

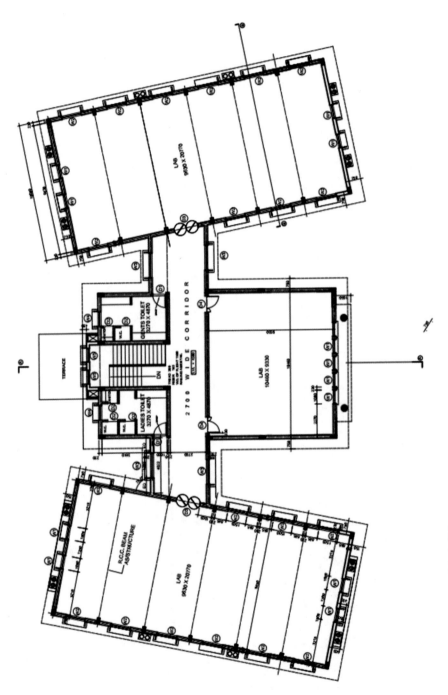

Fig. 5 Layout of first floor of the building

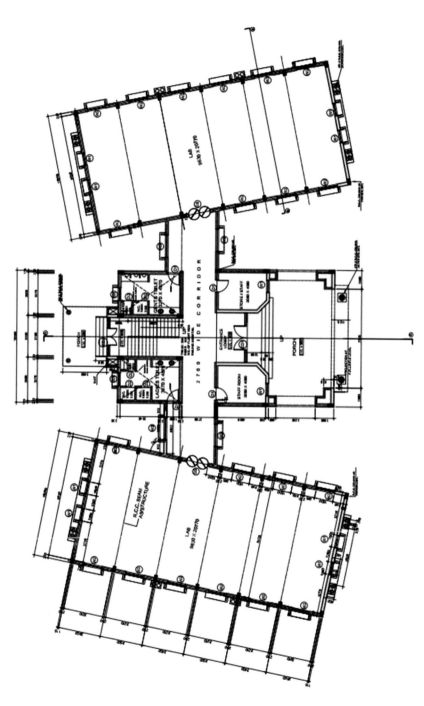

Fig. 6 Layout of ground floor of the building

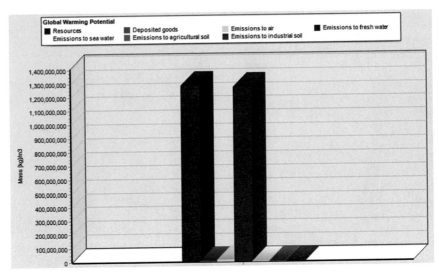

Fig. 7 Emissions associated with the building

$$Total\, electrical\, consumption = Average\, monthy\, electrical\, consumption \times$$
$$No.\, of\, months\, in\, a\, year \times Building\, Life\, Cycle\, (Years)$$
$$(1)$$

In the construction phase, different types of materials like cement, sand, steel etc. are extensively used. But in the present case study cement, sand and bricks are considered only and rest of the construction materials are excluded due to unavailability of ample data. A lot of primary energy is utilized from their extraction of raw materials, manufacturing, final processing and assembly; and transportation to the required destination [7]; Reddy 2004; [20, 22]. Along with this, GHG emissions are also associated with these processes related to the construction materials used in building. In order to estimate the carbon footprints associated with construction process of the building, all these GHG emissions must be taken into account. Maintenance and other repair activities are the important parts of a building and consume a lot of energy. This amount of energy utilized in the maintenance works is also varied with amount of raw materials used, its scale and type of method used. However, the present case study does not consider any maintenance and repair works. This is due to the fact that the selected building is only 9 years old and few small maintenance works have been carried out during this period and therefore excluded from the present study. In addition to this, unavailability of proper maintenance data is another issue, which forces to leave the maintenance phase in the present life cycle study. Another important phase of a building's life cycle is demolition phase. Demolition of a building requires certain amount of energy and a certain amount of emissions are associated with it. The amount of energy consumption and associated GHG emissions may vary with type of method and technology used to demolish the building. But demolition of a building is not a common process and it is a future and unpredictable task. Also, many buildings may go for reconstruction and modifications to upgrade its duration of its life cycle instead of demolition. In the present case study, demolition phase of the building is excluded and analysis is carried out with construction and operational phase of the building. All collected data related with construction and operational phase of the building are used to find the emissions associated with the selected building using GaBi Education software. Figure 7 shows the results in terms of emissions associated with the selected building.

5 Discussion

It can be noticed from the present analysis that buildings have considerable share in releasing carbon emissions into the atmosphere. The Global Warming Potential (GWP) related to various construction materials, primary energy input, electricity associated with the studied building is estimated. The total value of GWP comes out to be 1.59×10^6 kg-CO_{2eq}. Out of which the manufacturing of cement using in building construction has the biggest share with 6.63 kg-CO_{2eq}, which is then followed by thermal energy required in manufacturing of bricks. Figure 7 shows the emissions associate with the building in air, freshwater, sea water, agriculture soil and industrial soil. It can be seen by the figure that the building affects the freshwater the most with total emissions of 1.27×10^9 kg/m^3, which is then followed by emissions to air and emissions to sea water. These results indicate that the consumption of large quantity of resources, energy, operational electrical energy in a building have significant impacts to the atmosphere in terms of GHG emissions.

The results obtained from the present analysis are also compared with other previous studies. Table 1 shows some studies available on LCA of educational buildings carried out by various researchers summarized with the present study. Scheuer et al. estimated emissions associated with a university building per unit floor area [22]. Arena and Rosa analyzed impacts associated with a school building in Argentina in terms of GWP, photo smog, resources consumption etc. per unit area of considered refurbishment technique [1]. Ozawa-Meida et al. estimated carbon footprint for a university building in UK considering internal area without construction [18]. In 2012, Varun et al. studied an academic building and calculated GHG emissions per unit usable floor area [29]. Isasa et al. estimated GHG emissions for a building in Spain/Portugal per unit acclimatized area for use phase [8]. In fact, it is difficult to compare all these studies as every study has different scope and boundary conditions. There are many variations in the methodologies and functional unit used in these studies. Many of them have not used any software to perform the LCA. As LCA may be seen as a complex and time consuming technique as a lot of data collections, calculations and their processing is to be done. Advent of software to perform LCA makes it simpler and realistic but along with this proper data and record keeping is quite essential to obtain more useful results. Due to lack of ample details of materials used in construction process, unavailability of on-site data is a big problem during the

Table 1 Comparison of life cycle GHG emissions with previous studies

S. No.	Year	Building specification	Country	GHG emissions
1	2003 [22]	Sam Wyly Hall, University	USA	67,500 ton
2	2003 [1]	School building	Argentina	34,000 μPE*
3	2011 [18]	University Building	UK	65 kg CO_2eq/m^2
4	2012 [29]	Academic Institute	India	0.45×10^3 kg CO_2eq/m^2 usable floor Area
5	2014 [8]	–	Spain/Portugal	514 MJ/m^2 acclimatized area
6	2012 [16]	Academic Institute	India	978,873.40 kg-CO2eq/m^2 usable floor area
7	Present study	Academic Institute	India	2.57×10^9 kg/m^3

*μPE: 10^6 person equivalents

inventory analysis. Due to these, many important data have to be excluded and this may lead to the results with greater uncertainty. Nevertheless, these results obtained from LCA studies cannot be ignored since all these studies estimate considerable amount of emissions into the atmosphere. As discussed that there may be some level of uncertainty in the results obtained from LCA of buildings because of not considering many factors/inputs–outputs but the fact must be considered that the actual emissions must be higher than these emissions. Therefore, it becomes quite important to adopt those materials and techniques, which make the construction and operation of buildings more sustainable and less harmful to the environment.

6 Conclusion

Buildings have significant share in total global GHG emissions as they consume a large variety of inputs and energy throughout their complete life cycle which are responsible for releasing high amount of emissions into the atmosphere. In order to mitigate these emissions and move forward to sustainable development, it is important to explore those inputs in construction of a building, its operation and maintenance, demolition which consumes lot of energy and release emissions. LCA is a useful and promising technique to study environmental performance of building throughout its life cycle and estimate the environmental degradation associated with it in terms of various environmental performance indicators. In the present study, LCA is carried out on an academic building located in a hilly region and emissions

associated with it have been quantified. Location of a building is an important parameter, which affects the value of associated emissions due to increase in transportation
distance. It is also found that academic buildings have lower resources consumption
and associated emissions than those related to residential buildings due to less or only
day time operational hours. Another important thing to perform an effective LCA is
availability of data of all resources utilized by a building especially in construction
and operational phase. In the present study, unavailability of ample data is faced due
to which many important inputs had to be excluded. Nevertheless, the results obtained
in the study clearly reveal that emissions associated with the building have notice-
able effect on the natural resources and it is important to explore new sustainable
construction techniques and materials; and formulation of new strategies to mitigate
these effects.

Acknowledgments The authors are grateful to the authorities of THDC Institute of Hydropower
Engineering & Technology, Tehri (India) for allowing and providing valuable data required to carry
out the research work.

References

1. Arena AP, Rosa C (2003) Life cycle assessment of energy and environmental implications of
   the implementation of conservative technologies in school building in Mendoza-Argentina.
   Build Environ 38:359–368
2. Basket JS, Lacke CJ, Weizt KA, Warren JL (1995) Guidelines for assessing the quality of life
   cycle inventory analysis. Research Triangle Institute, National Technical Information Service,
   Springfield
3. Central Electricity Authority (CEA) (2020) Ministry of power government of India. (www.
   cea.nic.in)
4. Dakwale VA, Ralegaonkar RV, Mandavgane S (2011) Improving environmental performance
   of building through increased energy efficiency: a review. Sustain Cities Soc 1:211–218
5. Dhaka S, Mathur J, Garg V (2012) Combined effect of energy effciency measures and thermal
   adaptation on air conditioned building in warm climatic conditions of India. Energy Build
   55:351–360
6. Gamarra AR, Istrate IR, Herrera I, Lago C, Lizana J, Lechon (2018) Energy and water consumption and carbon footprint of school buildings in hot climate conditions. Results from life cycle
   assessment. J Cleaner Prod
7. Gumaste KS (2006) Embodied energy computations in buildings. Adv Build Res 404–409
8. Isasa M, Gazulla C, Zabalza I, Oregi X, Partidario P, Duclos L (2014) EnerBuiLCA: life cycle
   assessment for energy efficiency in buildings.Conference: WSB14 Barcelona, 124, 1–10
9. ISO, ISO 14040 (1997) Environmental management—life cycle assessment—principles and
   framework. International Organisation for Standardization
10. Khasreen MM, Banfill PFG, Menzies GF (2009) Life-cycle assessment and the environmental
    impact of buildings: a review. Sustainability 1:674–701
11. Levine MD, Price L, Martin N (1996) Mitigation options for carbon dioxide emissions from
    buildings. Energy Policy 24(10/11):937–949
12. Magrassi F, BorghiAD Gallo M, Strazza C, Robba M (2016) Optimal planning of sustainable
    buildings: integration of life cycle assessment and optimization in a decision support system
    (DSS). Energies 9:1–15

13. Nautiyal H, Varun (2015) CDM: a key to sustainable development. In: Thangavel P, Sridevi G (ed) Environmental sustainability—role of Green Technologies, 121–128
14. Nautiyal H, Shree V, Khurana S, Kumar N, Varun (2015) Recycling potential of building materials: a review.Muthu SS (ed) Environmental Implications of Recycling and Recycled Products, 31–50
15. Nautiyal H, Shree V, Singh P, Khurana S, Goel V (2018) Life cycle assessment of an academic building: a case study. Muthu SS (ed) Environmental carbon footprints, 295–315
16. Nautiyal H, Varun (2012) Progress in renewable energy under clean development mechanism in India. Renew Sustain Energy Rev 16, 2913–2919
17. Nautiyal H, Goel V, Singh P (2019) Water footprints of hydropower projects. Muthu SS (ed) Environmental water footprints (energy and building sectors), 35–46
18. Ozawa-Meida L, Brockway P, Letten K, Davies J, Fleming PD (2011) Measuring carbon performance in a UK University through a consumption-based carbon footprint: De Montfort University case study. J Cleaner Prod 5:185–198
19. Ramesh T, Prakash R, Shukla KK (2012) Life cycle energy analysis of a residential building with different envelopes and climates in Indian context. Appl Energy 89:193–202
20. Reddy BVV, Jagadish KS (2003) Embodied energy of common and alternative building materials and technologies. Energy Build 35:129–137
21. Sartori I, Hestnes AG (2007) Energy use in the life cycle of conventional and low energy building: a review article. Energy and Build 39:249–257
22. Scheuer C, Keoleian GA, Reppe P (2003) Life cycle energy and environmental performance of new university building: modelling challenges and design implications. Energy Build 35:1049–1064
23. Sharma A, Saxena A, Sethi M, Shree V, Varun (2011) Life cycle assessment of buildings: a review. Renew Sustain Energy Rev 15, 871–875
24. Shree V, Varun, Nautiyal H (2015) Carbon footprint estimation from a building sector in India. In: The carbon foot print handbook, 239–258
25. Srinivasan RS, Ingwersen W, Trucco C, Ries R, Campbell D (2014) Comparison of energy-based indicators used in life cycle assessment tools for buildings. Build Environ 79:138–151
26. Tsai WH, Lin SJ, Liu JY, Lin WR, Lee KC (2011) Incorporating life cycle assessments into building project decision-making: an energy consumption and CO_2 emission perspective. Energy 36:3022–3029
27. Varun Bhat IK, Prakash R (2009) LCA of renewable energy for electricity generation systems e a review. Renew Sustain Energy Rev 13(5):1067–1073
28. Varun PR, Bhat IK (2010) Life cycle energy and GHG analysis of hydro electric power development in India. Int J Green Energy 7(4):361–375
29. Varun Sharma A, Shree V, Nautiyal H (2012) Life cycle environmental assessment of an educational building in Northern India: a case study. Sustain Cities Soc 4:22–28
30. Varun, SA, Nautiyal H (2016) Environmental impacts of packaging materials, environmental footprints of packaging. Muthu SS (ed) Environmental footprints of packaging (environmental footprints and eco-design of products and processes), 115–137
31. Wackernagel M, Rees WE (1996) Our ecological footprint: reducing human impact on the earth
32. Wiedmann T, Minx J (2008) A definition of "carbon footprint." In: Pertsova CC (ed) Ecological economics research trends, 1, 1–11
33. World Resource Institute (www.wri.org)

Toward a Low-Carbon Economy: The Clim'Foot Project Approach for the Organization's Carbon Footprint

Simona Scalbi, Patrizia Buttol, Arianna Dominici Loprieno, Gioia Garavini, Erika Mancuso, Francesca Reale, and Alessandra Zamagni

Abstract The EU Emission Trading System (ETS) represents an essential part of the European policies on Climate Change, targeting the most polluting organizations, which cover 45% of the GHG emissions. However, no common framework has been proposed yet for "non-ETS organizations." The reduction of direct emissions in most of the cases is not enough for significantly tackling climate change, but an approach that encompasses also indirect emissions should be adopted, as promoted in the Carbon Footprint of Organisations (CFO), for achieving the ambitious targets set in the European Green Deal. The application of the CFO supports organizations in defining and monitoring the effects of mitigation actions: thanks to CFO, organizations are encouraged to innovate their management system, improve the use of resources, strengthen relationships in the supply chain, beside obtaining a reduction of their costs. In this context, the LIFE Clim'Foot project has given a contribution to foster public policies for calculation and reduction of the CFO. The project has dealt with two key aspects: (i) the need for national policies addressing GHG emissions of non-ETS organizations and the strategic role of structured and robust tools, such as national databases of Emission Factors; (ii) the relevance of organizations'

S. Scalbi (✉) · P. Buttol · A. Dominici Loprieno · E. Mancuso
Laboratory Resources Valorisation, ENEA, Rome, Italy
e-mail: simona.scalbi@enea.it

P. Buttol
e-mail: patrizia.buttol@enea.it

A. Dominici Loprieno
e-mail: arianna.dominici@enea.it

E. Mancuso
e-mail: erika.mancuso@enea.it

G. Garavini · F. Reale · A. Zamagni
Ecoinnovazione s.r.l., Via Ferrarese 3, 40128 Bologna, Italy
e-mail: g.garavini@ecoinnovazione.it

F. Reale
e-mail: f.reale@ecoinnovazione.it

A. Zamagni
e-mail: a.zamagni@ecoinnovazione.it

© The Author(s), under exclusive license to Springer Nature Singapore Pte Ltd. 2021
S. S. Muthu (ed.), *LCA Based Carbon Footprint Assessment*, Environmental Footprints and Eco-design of Products and Processes, https://doi.org/10.1007/978-981-33-4373-3_4

training in fostering their commitment to account for and mitigate GHG emissions. This chapter illustrates the development and application of Clim'Foot approach for promoting the calculation of the CFO and definition of mitigation actions and to highlight the results of the testing phase in Italy. The approach is described in terms of (i) the toolbox developed (national databases of emission factors, training materials and carbon footprint calculator), (ii) the voluntary program set up to engage public and private organizations and (iii) the role played by decision-makers. Strengths and weaknesses of the Clim'Foot approach are discussed, together with opportunities of replicability and transferability of the results to support the development of a dynamic European network for carbon accounting.

Keywords Climate change · Carbon footprint · CFO · Low carbon economy · Mitigation actions · Emission factors · Data quality · Carbon footprint calculator

1 Introduction

Human activities, especially combustion of fossil fuels, deforestation and farming livestock, have led to an increase of the global average temperature for about 0.85 °C in the last 20 years [1]. Scientists consider that an increase of 2 °C compared with preindustrial age is the threshold, beyond which we can expect dangerous and even catastrophic event occurring. For this reason, 195 countries reached an agreement at the Paris Climate Conference (COP21), held in December 2015, to limit global warming to well below 2 °C above preindustrial levels. Actually, the national climate action plans presented in Paris showed just the trend to be followed but are not enough to achieve the goal.

Also before 2015, the EU countries together with Iceland had endorsed the Kyoto protocol (1998) and were committed to cut by 20% compared to 1990 the greenhouse gas (GHG) emissions by 2020. Moreover, the EU has defined a road map of the transformation towards a low-carbon economy [2], which engages the EU to achieve 40% reduction of GHG emissions by 2030, compared to 1990, and 80% by 2050.

In this context, the EU emission trading system (ETS) represents an essential part of the European policies on climate change, as it targets the most polluting organizations, which cover 45% of the GHG emissions. The ETS sectors that mostly contribute to GHG emissions are the following[1]:

- power and heat generation, all energy-intensive industry sectors (oil refineries, steel works and production of metals, cement, lime, glass, ceramics, pulp, paper, cardboard, acids and organic chemicals), commercial aviation, which are mainly sources of carbon dioxide (CO_2);
- production of nitric, adipic, glyoxal and glyoxylic acids as sources of Nitrous oxide (N_2O);
- aluminum production as a source of perfluorocarbons (PFCs).

---

[1] http://ec.europa.eu/clima/policies/ets/index_en.htm.

In 2014, the European Council confirmed that emission reductions should be reached not only by the ETS but also by non-ETS sectors, setting the reduction targets to 43% and 30%, respectively, by 2030 compared to 2005 [3].

To reach this commitment, the Effort Sharing Regulation[2] has established binding annual GHG emission targets for Member States for the periods 2013–2020 and 2021–2030. These targets concern emissions from most non-ETS sectors, such as transport, buildings, agriculture and waste, which can led to a reduction of the total EU emissions by 10% [4]. Currently, 13 Member States have reached reduction of GHG emissions, but only 4 have already fulfilled their 2020 goals. To reach their targets, the States need to implement additional measures (already in place in some countries), e.g. energy efficiency measures in the residential and services sectors, or to use the flexibility mechanisms that Effort Sharing Regulation makes available [5].

However, no common framework has been proposed yet for GHG reduction targets at country level for non-ETS organizations.[3] Consequently, the involvement of private and public organizations to reduce their carbon footprint (CF) should be supported by reliable data, tools and innovative approaches that allow also the calculation of their GHG emissions. Besides, the environmental advantages due to the mitigation actions identified are accompanied by innovation, optimization of the resources use, strengthening of relationships within the supply chain and reduction of management costs.

In this context, the LIFE project "Clim'Foot—Climate Governance: Implementing public policies to calculate and reduce organizations carbon footprint" (LIFE14 GIC/FR/000475, hereinafter Clim'Foot[4]) aimed to foster public policies for calculation and reduction of the CF of non-ETS organizations. The project has dealt with two key aspects:

(i)   the need for national policies addressing GHG emissions of non-ETS organizations and the strategic role of standardized tools, such as national databases (DBs) of Emission Factors (EFs)[5]

(ii) the relevance of organizations training to foster their commitment targets to account and mitigate GHG emissions.

Clim'Foot has brought together seven partners from five EU Countries: ADEME (project coordinator) and IFC (France), ENEA and Ecoinnovazione (Italy), CRES (Greece), HOI (Hungary) and EIHP (Croatia).

After an overview of the Standards for CFO calculation, development and application of the Clim'Foot approach for calculating and reducing CFO are here presented and discussed, together with the results achieved from its implementation in public and private organizations in Italy.

---

[2]https://ec.europa.eu/clima/policies/effort/regulation_en.

[3]http://ec.europa.eu/clima/policies/strategies/progress/kyoto_2/index_en.htm.

[4]https://www.climfoot-project.eu/.

[5]The Emission Factors are calculated ratios between the quantity of GHG emissions and the units of activity associated with their release [6].

2 Standards for Carbon Footprint Organization

In literature, standards and specifications exist to calculate the CFO, such as the GHG Protocol Corporate Accounting and Reporting Standard [6]; the GHG Protocol Corporate Value Chain [7] and the ISO 14064 [8], Part 1.

These documents provide organizations with directions about identifying, measuring and communicating the GHGs emitted, generally in 1 year, from all the activities (direct and indirect) across the organization, including the use of energy in buildings, industrial processes and company vehicles.

ISO and GHG Protocols present a similar approach to the calculation of the CFO. The former, like all the standards, provides the reference framework, while the latter goes into the detail of its implementation and contains also motivations for GHG reporting [9].

Both documents propose two types of approaches for setting organization's boundaries:

- Approach of control: all GHG emissions and/or removals are quantified concerning facilities that the organization controls financially or operationally. A company has financial control over the operation if it has the right to the majority of benefits of the operation or if it retains the majority risks and rewards of ownership of the operation's assets. A company has the operational control if it has the authority to introduce and implement its operating policies [6] [page 17].
- Equity share approach: the organization quantifies its portion of GHG emissions and removals from respective facilities [8]. The equity share reflects the extent of rights a company has to the risks and rewards coming from an operation and is normally the same as the ownership percentage [6] [page 17].

The GHG protocols and ISO recommend the classification of three types of emissions:

1. *Direct GHG emissions*: emissions from greenhouse gas sources owned or controlled by the company, defined as *Scope 1* by GHG Protocol.
2. *Energy indirect GHG emissions*: emissions from the production of purchased energy used by the company (electricity, heat or steam), defined as *Scope 2* by GHG Protocol.
3. *Other indirect GHG emissions,* e.g., emissions from business travel by employees, transport of products and materials, waste generated by the organization but managed by another organization, defined as *Scope 3* by GHG Protocol.

Under Scope 1 the following emissions are considered:

- Emissions from fuels and/or Wastes burning.
- Process and Fugitive emissions from:

  - Air conditioning and cooling
  - Agriculture

- Industrial process
- Wastes

- LULUCF[6] (Land use, Land Use Change and Forestry).

Scope 2, which includes emissions from the production of the purchased energy used by the Organization, does not include the transmission and distribution losses, which are accounted for in scope 3. By definition, scope 3 emissions are all indirect emissions (not included in scope 2) occurring in the value chain (e.g., materials suppliers, third-party logistics providers, waste management suppliers, travel suppliers, employees, and customers) [6]. The choice about the categories to be included in scope 3 is discretionary and this may impair comparison across companies.

Scope 3 accounts also GHG emissions of capital goods (i.e., plant, property and equipment, such as furniture, office equipment, and computers that the company uses for its activity). Since these GHG emissions are not depreciated or discounted over the life of the asset, which typically occurs in financial accounting, capital purchases, such as new building construction, occurring only once in a while, may significantly vary scope 3 emissions from year to year and companies should highlight the exceptionality of the capital investment in the public report [7].

Moreover, the scope 3 accounting is based on the life cycle approach and carbon footprint is one specific indicator accounted for in LCA, so the CFO method including the scope 3 is a starting point to become familiar with Life Cycle Assessment.

In literature, there are several resources for carbon calculation and carbon disclosure options available for businesses, institutions and local authorities. For microorganizations and SMEs, many of these resources are free. Table 1 reports some examples of free calculators, all including national emission factors.

3 The Clim'Foot Approach to Carbon Footprint of Organizations

The Clim'Foot approach for CFO calculation and reduction is an original concept, developed and tested during the project, which is structured along three levels (Fig. 1):

(a) development of a toolbox including national DBs of EFs, a tool for the calculation of CFO, training materials and a dissemination platform;
(b) setting up of a voluntary program, involving a selected number of proactive public and private organizations, which are trained for using the toolbox to calculate their CFO, with the support of technical experts;
(c) involvement of policymakers since the early stage of the process, to foster the replicability and transferability of the approach and the implementation of regulations or public policies for the mitigation of GHG emissions.

---

[6]The LULUCF covers emissions of GHG and removal of carbon from the atmosphere due to human use of soils, trees, plants, biomass and timber.

Table 1 Examples of CFO free calculators

Source	Description	Web site
The Department for Business Energy & Industrial Strategy (BEIS) of UK	They provide a tool (Excel format) and a guide [10]	https://www.gov.uk/government/publications/greenhouse-gas-reporting-conversion-factors-2018
The Environmental Protection Agency (EPA) of Ireland	They provide a list of carbon calculators and a tool for resource efficiency	http://www.BeGreen.ie and www.GreenBusiness.ie
Carbon Footprint Ltd	They propose an online carbon footprint calculator. They also offer other advanced tools for businesses for a fee	http://www.carbonfootprint.com/calculator1.html
GHG Protocol	They provide a suite of calculation tools to assist companies in calculating their GHG emissions and measuring the benefits of mitigation projects	https://ghgprotocol.org/calculation-tools

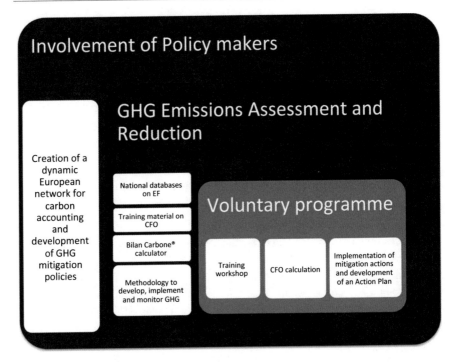

Fig. 1 Clim'Foot approach

3.1 The Toolbox

3.1.1 The National Databases

The national DBs (one per each of the countries involved in the project) include a set of European EFs, which are common to all countries, and a number of country-specific EFs. The databases are supplemented by technical documentation aiming to:

(i) share the data sources used and promote the replicability in other sectors and contexts;
(ii) enable validation and update of the EFs;
(iii) present the data to external users such as regulators, the public and stakeholder groups in a transparent way.

The documentation also serves the purpose of ensuring consistency among the different DBs in terms of completeness of data description, appropriateness of calculation and coherence of data quality assessment.

The development of national DBs on EFs requires a common methodology, which defines the content and classification structure of the DBs and the list of greenhouse gases with their characterization factors, gives recommendations on data collection and data quality requirements, overviews the main data sources available and ensures a consistent development of datasets starting from different data sources (e.g. Life Cycle Inventories (LCI) from LCA databases and National Inventory Reports). This is a major issue as some data sources deliver subsets of most relevant emissions, while others give the results in terms of CO_{2eq}, after aggregating the emissions according to the characterization factors.

The methodology adopted in Clim'Foot [11] is mainly based on the GHG protocol for Organizations [6, 7, 12], the ISO 14064 [8] and the IPCC guidelines [13], but takes also into account the European initiative on Organization Environmental Footprint (OEF) [14], in particular for the data quality definition.

The Clim'Foot DBs cover the sectors that most contribute to the GHG emissions following the recommendations of the GHG protocols, thus including Metals, Chemicals, Minerals, Pulp and paper, Semiconductor productions, Refrigerants, and the emissions related to land use, land use change and forestry (LULUCF).

In the DB, each emission factor is a record, and each record represents a unit process of human activity that exchanges resources (biogenic CO2 uptake) and GHG emissions with the environment. **Each activity represents a process/good/service** and is characterized by a specific **reference flow**, i.e., the measure of the **process/good/service** output. Each record includes:

- **metadata**, which qualitatively and quantitatively describe the emission factor to support the end user's choice of EF for the CF calculation;
- **elementary flows**, i.e. all the GHGs exchanged with the environment during the human activity;

- **characterized GHG** in CO_{2eq}, calculated by multiplying the amount of each GHG by its characterization factor
- **emission factor**, which is obtained by adding all characterized GHGs related to the human activity (mass unit of CO_{2eq}/amount of activity—e.g., $kgCO_{2eq}$/1 kg CH_4 production).

The elementary flows are the GHGs listed in the Kyoto Protocol and the characterization factors of IPCC 2013 are used (Table 2).

The robustness of each EF is evaluated by means of data quality criteria, which intend to answer two different questions: (i) how much does the EF represent the declared characteristics of the data sources from which it has been elaborated? (ii) how much suitable is the EF factor to assess the CFO of a specific company, i.e., how much does it fit for purpose? Building upon international and European initiatives on data quality in environmental footprint studies, the methodology provides instructions for the data quality evaluation by defining the following criteria:

- *time-related representativeness (TiR)* = "degree to which the dataset reflects the true population of interest regarding the time/age of the data, including for included background process datasets, if any" [15].
- *technological representativeness (TeR)* = "degree to which the dataset reflects the true population of interest regarding technology, including for included background process datasets, if any" [15].
- *geographical representativeness (GeR)* = "degree to which the dataset reflects the true population of interest regarding geography, including for included background process datasets, if any" [15].

A qualitative approach was chosen for its evaluation, taking into account the information available on the used data sources. Table 3 reports the description of the quality criteria used in the Clim'Foot DB.

Table 2 Characterization factors from IPCC 2013 [1]

Gases common name	Chemical formula	Characterization Factor in CO_{2eq}
Fossil Carbon dioxide (CO_2)	CO_2	1
Biogenic Carbon dioxide (CO_2)	CO_2	–
Methane	CH_4	30
Biogenic methane	CH_4	28
Nitrous oxide	N_2O	265
Hydrofluorocarbons[a]	HFCs	–
Perfluorocarbons[a]	PFCs	–
Sulfur hexafluoride	SF_6	23500
Nitrogen trifluoride[b]	NF_3	16100

[a]See Appendix 8.A of IPCC 2013 document for the complete list
[b]Nitrogen trifluoride (NF3) has been recently added to the requirements of Scope 3 Standard and Product Standard

Table 3 Quality level and rating for the quality criteria adopted in the Clim'Foot project

Quality level	TiR	TeR	GR
Very good	The TiR is not older than 4 years with respect to the reference year of the data source	The technologies used are exactly the same as the technologies covered by the data	The process takes place in the same country as the one the data is valid for
Good	The TiR is not older than 6 years with respect to the reference year of the data source	The technologies used are included in the mix of technologies covered by the data	The process takes place in the geographical region (e.g. Europe) for which the data is valid for
Fair	The TiR is not older than 8 years with respect to the reference year of the data source	The technologies used are similar to those covered by the data	The process takes place in one of the geographical regions for which the data are valid for
Poor	The TiR is not older than 10 years with respect to the reference year of the data source	The technologies used show several relevant differences compared with the technologies covered by the data	The process takes place in a country that is not included in the geographical region(s) the data are valid for, but sufficient similarities are estimated based on expert judgment
Very poor	The TiR is older than 10 years with respect to the reference year of the data source	The technologies used are not representative for the technologies covered by the data	The process takes place in a different country than the one for which the data are valid for

During the project, five national databases were developed in excel format (one file per country, Fig. 2). Each national DB file includes the following six sheets:

1. **Category**: it includes the categories for each language;
2. **National DB**: it includes the description of the metadata, the quantified emissions (by gas type) associated to each activity included in the DB, the associated EF and the EF breakdown per gas;
3. **Clim'Foot DB**: it includes all the National DBs developed in the project, including both country-specific and EU EFs;
4. **CHF** includes the Characterization Factors of HFCs;
5. **PFC** includes the Characterization Factors of PFCs;
6. **GHG** includes the Characterization Factors of CO_2, CH_4f, CH_4b, N_2O, SF_6.

Each national DBs has 150 European EFs, common to all databases, and at least 150 country-specific EFs. Table 4 shows the number of National and European EFs developed in the Clim'Foot project for each sector and country.

The extensive excel databases have been made available to the potential final users through a simplified web version, available on the Clim'Foot website.[7] Two types of search functions have been provided: within a specific category, i.e. the sectors

---

[7] http://www.climfoot-project.eu/.

Process Name (English Language)	ClimF OOT ID	Copyrights	Data collector	Source	Unit (English language)	Technical description (English language)	Aggregated EF	Unit of EF in National language
Fuel oil - L (IT)	IT00202	ClimFoot project	ENEA	Italian National Inventory Report 2016	L	The main information available nationally of fuel oil EF is a sizable difference in carbon content between high sulphur and light sulphur brands. The data were elaborated from literature and from an extensive series of samples (more than 400) analysed by ENEL and made available to ISPRA. Carbon content varies to a certain extent also between the medium sulphur content and the very low sulphur products, but the main discrepancies refer to the high sulphur type. According to the available statistical data, it was possible to trace back to the year 1990 the produced and imported quantities of fuel oil divided between high and low sulphur products and to estimate the average carbon emission factor (NIR, 2016)	2,64E+00 kgCO2e/L	
LPG - L (IT)	IT00203	ClimFoot project	ENEA	Italian National Inventory Report 2016	L	The data set includes the data elaborate by ISPRA with the purpose to analyse regularly the chemical composition of the used fuel or relevant commercial statistics to estimate the carbon content / emission factor (EF) of the fuels. ISPRA has made investigations on the carbon content of the main transportation fuels sold in Italy, petrol, diesel and LPG, with the aim of testing the average fuels from 2012 to 2014. The goal of work was the verification of CO2 emission factors of Italian energy system, with a particular focus on the transportation sector.	1,54E+00 kgCO2e/L	
bottle water in PET 0,5 L (IT)	IT00204	ClimFoot project	ENEA	Fantin V., S. Scalbi, G. Ottaviano, P. Masoni, (2014) "A method for improving reliability and relevance of LCA reviews: The case of life-cycle greenhouse gas emissions of tap and bottled water", Science of The Total Environment, Volumes 476–477, 1 April 2014, Pages 228–241, DOI:10.1016/j.scitotenv.2013.12.115.	L	0,5 l PET bottle of water, produced in Italy, includes production and distribution. Boundaries from the cradle to the gate. Average data obtained from the article, Fantin V., S. Scalbi, G. Ottaviano, P. Masoni, (2014) "A method for improving reliability and relevance of LCA reviews: The case of life-cycle greenhouse gas emissions of tap and bottled water", Science of The Total Environment, Volumes 476–477, 1 April 2014, Pages 228–241, DOI:10.1016/j.scitotenv.2013.12.115.	2,07E-01 kgCO2e/L	
Cookies	IT00214	ClimFoot project	ENEA	Petri/ Nicoletta, Marchettini Nadia, Niccolucci Valentina, Pulselli Federico M., 2018, Steps towards SDG 4: teaching sustainability through LCA of food, Proceedings of the 12th Italian LCA Network Conference Messina "Life Cycle Thinking in decision-making for sustainability: from public policies to private businesses", 11-12th June 2018 Edited by Giovanni Mondello, Marina Mistretta, Roberta Salomone, Arianna Dominici Loprieno, Sara Cortesi, Erika Mancuso, ISBN: 978-88-8286-372-2	kg	Packaged cookies included the production and distribution. Boundaries from the cradle to the gate. Average data from 11 Italian EPD studies	1,50E+00 kgCO2eq/kg	

Fig. 2 Database format, sheet on National DB

Table 4 National and European emission factors developed in the Clim'Foot project for each sector and country

Sector		European	Hungarian	Croatian	Greek	Italian	French
Energy	Fuels	23	19	18	25	43	67
	Electricity	6	4	1	26	4	
	Thermal energy	6	13	21	20		
Fugitive emissions				9 (refrigerant)		29 (agriculture)	
Transport (freight and passenger)	Road	3	250	91	22	73	89
	Rail	2	32	4	2		
	Air	5		4	8		
	Water	3		2	4		
Industrial Processes and Product use	Materials	59	34	5	9		
	Chemicals	21				9	
	Construction	12	13				
	Food and meals		1		27	16	
	Agriculture			5	10	14	
Waste Management		10	11	6	8	10	
LULUCF			6	6	9		
TOT		150	383	172	170	198	156

identify by the project, or among all the categories using three filters: keyword, localization or unit of emission factor (Fig. 3).

In the web version for each EF, a short description of metadata is given in two languages: the country-specific language of the DB and English (Fig. 4).

Fig. 3 Search format of the DB web version

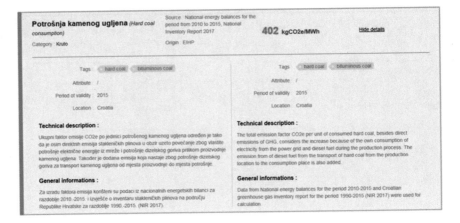

Fig. 4 Example of metadata in the DB

3.1.2 Carbon Footprint Organization Calculation Tool

In the context of the Clim'Foot project, the Bilan Carbone® Clim'Foot tool has been developed and used for the voluntary program. The tool has been developed starting from the French tool "Bilan Carbone®" by ADEME, in collaboration with the Association Bilan Carbone (ABC), and is currently managed by ABC. More in detail, the Bilan Carbone® has been adapted to the national contexts of the Clim'Foot partners through the integration of the national and European EF databases developed during the project.

The Bilan Carbone® Clim'Foot allows the GHG accounting in compliance with the GHG Protocol [6] and ISO standard 14064-1 [8] and provides formats for reporting accordingly. It consists of an Excel file including the following spreadsheets:

- A first spreadsheet, where to describe the organization performing the CFO and to select the approach to the CFO calculation (financial control, operational control, share of capital)
- Several spreadsheets to fill in the main activity categories described below
- A spreadsheet reporting all emission factors available for the calculation
- A supporting spreadsheet including references for conversion between several units of measure describing activity data
- Four spreadsheets reporting results in terms of graphs, CO2eq general overview, summary results tables compliant to ISO standard and to the GHG Protocol.

The Bilan Carbone® Clim'Foot tool allows GHG emissions calculation for all activities considered in the value chain of an organization, structured into 10 main categories:

- Energy sources, where it is possible to account for emissions from heat and electricity consumption

- Nonenergy related sources, where direct emissions can be calculated (e.g. from livestock or use of fertilizer) or directly reported. Such category addresses also gases not covered by the Kyoto protocol
- Inputs, where impacts of materials and products consumption in the organization operations are addressed (including water)
- Packaging, which accounts for emissions due to the production of packaging materials (e.g., used for the packaging of the finished products)
- Transport of people, where transport activities are commuting, business travels and external person travels (e.g. visitors)
- Transport of goods, which accounts for emissions due to upstream transport (purchasing), internal transport and downstream transport (selling, distribution)
- Waste, which accounts for emissions from the disposal/treatment of waste produced by the organization, during its production/service activity
- Capital goods, where emissions due to buildings and infrastructure (even IT) can be estimated, and the known or proper depreciation period is considered, if ISO standard 14064 is applied
- Use stages, which concerns emissions due to the use of products by consumers/final users
- End-of-life, which addresses emissions due to the disposal/final treatment of products, after their use.

The tool can be downloaded from the project website after registration and does not require high level of expertise even if, based on the experience of the voluntary program, the participation to the face to face training course has been judged to be very important. Several end users suggest developing a short guide with brief description of main methodology options and key features.

3.1.3 The Training Courses

The target groups of the training courses are trainers and end users. The training of the trainers has a twofold objective:

- to train trainers on the methodology for calculating and mitigating CFO;
- to learn coaching tips and strategies to be used for training the end users.

The objectives of the training of end users are: i) to increase awareness on the climate change impact; ii) to teach how to calculate the CFO and iii) to give an overview on how to plan and implement a carbon management plan.

Different training materials were made available:

- online training, providing a general overview on climate change, the methodology for the CFO assessment and the calculator use;
- material to prepare technical workshop on CF for organizations, in national languages;
- dissemination materials, to be used in meetings with stakeholders such as industrial trade associations or professional orders.

Fig. 5 Clim'Foot dissemination platform

The training course for organizations was implemented in the national languages of the Clim'Foot partners.

3.1.4 The Platform

The dissemination platform is a virtual place where users (policymakers and organizations) find information and training on climate change; have full access to the EFs national DBs and background reports; find information on the voluntary program already implemented, including the lessons learnt, and the description of how to implement a new voluntary program. The platform is available in six languages, namely English, French, Greek, Hungarian, Croatian, and Italian. The dissemination platform can be found at the Clim'Foot website (www.climfoot-project.eu/) (Fig. 5).

3.2 The Voluntary Program

The voluntary program includes three main phases: training, CFO calculation and reduction of the GHG emissions (i.e. implementation of mitigation actions). The program is based on a voluntary process to support public and private organizations in calculating and reducing their CF.

In the first phase, a large database of contacts was developed to identify end users interested in the project participation and a call for expression of interest was launched. Organizations that declared their interest signed an agreement for participating to the Clim'Foot project, and committed themselves to attend both on-line and two-day on-site training.

In the second phase, the organizations collected data and calculated their CFO. The layout of the CFO report is standardized and includes: information about the organization, definition of the boundaries, description of the data collection, results of the CFO calculation and hotspot analysis, which allows the identification of the main sources of GHG emissions. Most of the end-users reached a good understanding of the different aspects that have to be faced to calculate CFO, in particular how to select EFs for activities under their direct control (e.g. the consumption of purchased electricity) or relevant products/materials used in the production activities.

The third phase—reduction of the GHG emissions—is an important milestone of the Clim'Foot approach. Based on the hotspot analysis, a plan of mitigation actions has been defined—supported also by an economic evaluation—together with the definition of targets and a timeline. Examples of hotspots identified by public organizations were the reorganization and control of employees' travels, business travels and purchase of new company cars. The mitigation actions identified by organizations involved an increased use of renewable energy in combined heating and electricity system, the improvement of energy efficiency, the updating of heating systems. The actions for the transition to low-carbon organizations were classified into quick-wins (short term), priority actions (mid-term) and strategic actions, and their prioritization and the assessment of the results due to the action implementation were supported by a technical committee. Moreover, the communication of the carbon footprint results to several departments of organizations and to suppliers increased their environmental awareness and offered the opportunity to identify new possible mitigation actions.

3.3 The Policymakers' Involvement

The approach proposed suggests an early involvement of policymakers in Technical National Committees, as they are in charge of implementing regulations and public policies for the reduction of GHG emissions. Such an involvement has a twofold advantage: on the one side, they can access training sessions from the beginning of the process related to the approach and its objectives; on the other side they can give feedbacks on the process and suggest future initiatives of replicability and transferability, taking into considerations needs and wishes at policy level.

The final goal is to create a dynamic European network for carbon accounting, able to answer the following expectations:

- raising awareness among policymakers at national, regional and local level;
- exchanging best practices;
- fostering replicability and transferability of the experiences.

Indeed, one of the main purposes of the LIFE Clim'Foot project was to disseminate in European countries outside the project consortium, a common approach for calculating the CFO with standardized and high-quality databases and the project partners are committed to support replicability initiatives beyond the end of the project (see also Sect. 4.4).

4 Application of the Approach in the Italian Context

4.1 The Italian National Database of EFs

Data currently available to perform a CFO study (such as the EF Database by IPCC) have been mostly developed to fulfill the accounting duties set by the EU legislation at Member State level and for those organizations that contribute most to GHG emissions (i.e. the most polluting industries), but they do not match the needs of the totality of EU organizations. Several EFs are provided only with reference to an international scale (in terms of geographical representativeness), an aspect that raises a two-fold issue. On the policy side, the lack of national EFs limits the implementation of policies fostering the CFO accounting and reduction. On the organizations' side, the use of international EFs lowers the accuracy of the results of the CF and discourages mitigation actions. As a consequence of the current data availability, the CFO is mostly applied by those organizations able to afford the cost of a deep investigation on EFs, with the support of consultants and/or with the use of commercial DBs (e.g. LCI DB developed for Life Cycle Assessment).

The Italian DB includes 150 European EFs and 198 country-specific EFs (Table 5). The DB structure, currently in excel format, is designed to allow the import of data in a relational DB to improve its replicability and transferability.

4.2 Toward a Growing Use of CFO by the Organizations: The Voluntary Program for a Bottom-Up Process

The Italian database of contacts included 150 public and private organizations. The call launched to officially involve the potential end users into the project has received 19 expressions of interest to join the voluntary program. The Italian partners prepared and sent a survey to identify the key drivers for the participation of the 19 organizations to the voluntary program and to map their expectations. Public organizations showed interest in the project due to the opportunity to participate in collaborative networks, to exchange ideas and experiences toward a more environmentally friendly organization. For the private sectors, the project was a good opportunity to share experiences, and to learn more on climate change and its management to establish new potential business relationships and to increase competitiveness.

Table 5 Italian emission factors

Category	Data source	Number of EFs
Fossil fuels consumption	Italian National Inventory Report 2017 [16]	43
Electricity consumption	National report on renewable energy [17] and Italian National Inventory Report 2017 [16]	4
Freight transport	National database on transport, elaborated [18]	16
Passenger transport	National database on transport, elaborated [18]	57
Chemicals	Italian National Inventory Report 2017 [16]	9
Waste	Italian National Inventory Report 2017 [16]	10
Agriculture	Leap Database the Global Database of GHG emissions related to feed crops for the agricultural product, developed by FAO [19]	14
Mineral water	Fantin et al. [20]	2
Fugitive emission from agriculture	Italian National Inventory Report 2017 [13]	29
Food and beverage	Patrizi et al. [21]	14

The end-users have been involved in the following activities:

- Training sessions (on-line and on-site) on climate change and CF assessment;
- Calculation of their CFO by using Bilan Carbone® Clim'Foot, translated into Italian.
- Implementation of the mitigation actions based on the CFO results.

Two workshops were held in Italy for the organizations involved in the national voluntary program. They were organized in sessions of teaching and exercising and included: (i) a general overview of the main challenges related to climate change and energy, and the international and national initiatives on carbon footprint; (ii) a presentation of the methodological and standard principles and the main phases of a CF project; (iii) a technical presentation of the Bilan Carbone® calculator and practical exercises; (iv) the definition of mitigation actions for carbon reduction and the presentation of some case studies. All the developed material was made publicly available at the project website.

During the calculation phase, the support of the national partners has been guaranteed thanks to monthly contacts and technical face-to-face and virtual meetings. In particular, the organizations needed to be supported in the following steps: i) choice of the approach (equity share or control); ii) definition of the boundaries, i.e. activities and processes to be included in the CFO study, in line with the organizations strategic goals; iii) choice of the activity data to collect and EFs. The face-to-face meetings with the organizations, which were aimed at better involving the end-users and analyzing in detail the major difficulties encountered, were an important element of the experimentation phase.

All the 19 Italian organizations have calculated their CFO. Four organizations have analyzed only direct emissions and energy indirect emissions and focused on energy consumption. The others have investigated the indirect emissions too, such as materials in input, packaging, home-to-work transport, waste and capitals good, in order to get a more complete view of their carbon footprint and to select targets and strategies for a (potential) GHG reduction. A survey, which was aimed at monitoring the economic impacts of the project actions, has shown that most of the time (51%) for the CFO calculation was dedicated to the definition of the system boundary and the data collection, followed by the hours for data input in Bilan Carbone® Clim'Foot and analysis of the results (40%) and the hours for educational activity (9%).

During the voluntary program, strengths and weaknesses of Clim'Foot tools were also assessed. The organizations highlighted the need to have additional EFs to calculate their CFO, which were then built up and implemented in the DB (waste and water treatment, minerals water, renewable energies). Further development is in progress, which includes construction, chemicals and waste scenarios. All organizations highlighted that this experience increased their awareness on GHG problems. Some of them decided also to include the results of CFO in their quality management plan as an indicator to evaluate the efficacy of improvement actions related to energy.

After the calculation of their CFO, 13 organizations have defined mitigation actions, aimed at improving the mobility of the employees, increasing the share of renewable energy, improving the heating systems and overall reducing the emissions from purchased goods. Moreover, one organization developed a Sustainability Report that presents the policy and the actions to improve its social and environmental sustainability, including the environmental benefits obtained thanks to the implementation of the mitigation actions. As the monitoring of the total CO_2 emissions does not permit the assessment of the efficacy of the mitigation actions because emissions depend on the annual production, they defined the indicator "CO_2 emitted for unit of production," which catches the real environmental improvement better than the measurement of the total energy saving. At the end of the Clim'Foot project, 2 organizations have also developed an Action Plan where:

- a steering committee was set up to provide a governance of the action plan;
- the objective and targets of GHG reduction were defined;
- a set of mitigation actions was identified as well as strategies towards the defined objective and targets, in terms of finance and timeline;
- a list of environmental performance indicators was selected to check the outcomes of mitigation actions, such as CO_{2eq} per meter of tissue produced;

Furthermore, by the end of the project in August 2018, two organizations changed their electricity mix and chose a supplier who provides 100% renewable energy, implementing a mitigation action for CFO reduction, in agreement with their environmental policy.

4.3   The Involvement of Italian Policymakers

Policymakers from all country partners showed an interest toward the strategy adopted, i.e., the development of standard tools that can be either directly applied or used as a basis to create country specific tools. During the project, about 100 policymakers from countries different from those of the project partners have also been reached to present the project results and collecting information about the carbon policies of their countries. A survey and 2 webinars have been organized and 39 policymakers, coming from 9 countries, were trained on the Clim'Foot toolbox during a workshop organized at the end of the project final conference.

In Italy representatives of the Ministry of the Environment and the Ministry of Economic Development were involved in the national technical committee. During the periodic meetings, the toolbox and the voluntary program have been presented and have generated interest. In particular, the Italian Ministry of the Environment, signed a letter of commitment to endorse Clim'Foot tools, especially the Italian National Database and the Italian version of Bilan Carbone®. Indeed, the Clim'Foot initiative, with its tools and training courses, contributed to the project "CReIAMO PA" (Competences and Networks for Environmental Integration and Improvement of the Public Administration Bodies), carried out by the Italian Ministry of Environment and financed by the National Operational Programme on Governance and Institutional Capacity (Expertise and network for the environmental integration and for the improvement of organizations of the Public Administration 2018–2021). The project includes, among others, a Work Package—"Promotion of Environmental and Energy Management Models in Public Administrations," aimed at supporting local administrations that intend to plan and implement measures to reduce greenhouse gas emissions and improve the environmental performance of their organization.

The participation of the Città Metropolitana di Torino (Italy) is an example of involvement of a local public administration. After the training workshop, they contacted some schools of the territory and encouraged them to calculate and reduce their CFO. A group of students of five high schools were trained, calculated their schools CF and identified the main critical aspects. The participation to the voluntary program offered the public administration a twofold benefit: on one side, they could fulfill the demand for increasing environmental awareness of young people, in agreement with the objectives of the Green Education initiative of Piedmont Region; on the other side they could enrich the set of indicators monitored by the Energy manager of the Città Metropolitana, by adding the quantification of the schools CF.

4.4   The Post-life Phase

The update and the enlargement of the Italian national DB, which is time consuming and needs specific expertise, is another important aspect that is guaranteed after the end of the project by the commitment of the Clim'Foot partners through the collaboration with other projects aimed at developing country-specific datasets for both national EF and Life Cycle Assessment database, and through the involvement of stakeholders such as trade associations, national agencies or networks. The dissemination activities about the Clim'Foot approach are continuing in order to involve other policymakers and foster synergies with other projects or policymakers' initiatives. A first synergy has already been set between Clim'Foot and the project "CReIAMO PA," which is on-going and will run until June2022, and includes training and support actions on CFO.

From 2018 and so far, several training events were organized for the Public Administrations:

- five workshops in Rome, Turin, Cagliari, Padua and Bari to provide general training on CFO and present Clim'Foot tools
- five laboratory sessions in Florence, Turin, Bari, Cagliari and Padua for the use of the Clim'Foot calculator and to present CFO case studies.

Furthermore, two courses were organized for training the consultants that will support local administrations in calculating the CFO and in developing GHG mitigation actions. This training action is built upon the Clim'Foot action "training the trainers." Other training sessions are planned for the next future.

After the dissemination phase, the "CReIAMO PA" project started to calculate the CFO with several different public administrations: one Region (Piedmont), five Municipalities (Mantova, Prato, Cagliari, Saluzzo, Serrenti), three Regional agencies for environmental protection (Piedmont, Friuli Venezia Giulia, Apulia), and IPLA,[8] society for forestry management of Piedmont.

5   Conclusions

The modular structure of the toolbox and the integration of informative materials and documents summarizing the lessons learnt, are the strengths of the project. All this increases the potential of replicability and sustainability of the approach both in the consortium's countries and in other European countries. The ultimate goal is to raise policymakers' awareness on climate change mitigation and give them instruments, methodologies and training materials to implement the necessary national policies.

National Databases of GHG Emission Factors, with reliable and country-specific data available for free, are a decisive asset for national policies and can promote the adoption of mitigation actions by the organizations. The development of a common methodology and format, the definition of procedures for data collection, highlighting the main sectors to be developed and favoring the exchange of EFs among partners, were key aspects for the transferability and replicability of the approach in other countries and contexts.

---

[8]Istituto per le Piante da Legno e l'Ambiente.

The voluntary program has highlighted that training and support actions are important for the CFO calculation and the definition and implementation of mitigation policies. This experience has given useful feedback on the main tools developed in the project, namely the database and the training. For the database, the organizations identified the need to update some existing EFs, such as the Italian electricity mix and transportation, and to develop new EFs in some sectors such as chemicals and waste. Moreover, during the voluntary program, many organizations have requested further explanations about the system boundary definition and the difference among approaches (operational, financial). Therefore, the training structure was improved to include a more detailed explanation of these features and integrate new and more practical examples. This new structure has already been implemented in the training developed during the "CReIAMO PA" workshops and laboratories.

Acknowledgements The authors acknowledge the European Union LIFE Programme funding and all Clim'Foot partners for their contribution, support and collaboration in project activities. The authors are grateful to the Italian Ministry of Environment, who manages the CReIAMO PA project with the support of SOGESID (in house company of the Ministry), for the collaboration during the project.

References

1. IPCC (2013) Climate Change 2013: The Physical Science Basis. Contribution of Working Group I to the Fifth Assessment Report of the Intergovernmental Panel on Climate Change [Stocker TF, Qin D, Plattner G-K, Tignor M, Allen SK, Boschung J, Nauels A, Xia Y, Bex V, Midgley PM (eds). Cambridge University Press, Cambridge and New York, 1535 pp
2. EC (2011) European Commission, A Roadmap for moving to a competitive low carbon economy in 2050
3. EC (2014) The European economic and social committee and the committee of the regions—a policy framework for climate and energy in the period from 2020 to 2030, COM(2014) 15 final
4. EC (2016) European Commission, regulation of the european parliament and of the council: on binding annual greenhouse gas emission reductions by Member States from 2021 to 2030 for a resilient Energy Union and to meet commitments under the Paris Agreement and amending Regulation No 525/2013 of the European Parliament and the Council on a mechanism for monitoring and reporting greenhouse gas emissions and other information relevant to climate change, COM(2016) 482 final 2016/0231(COD)
5. Progress towards the 2020 Greenhouse Gas Target in Europe (2020) https://climatepolicyin fohub.eu/progress-towards-2020-greenhouse-gas-target-europe
6. World Resources Institute and World Business Council for Sustainable Development, GHG Protocol Corporate Accounting and Reporting Standard (2004)
7. World Resources Institute and World Business Council for Sustainable Development, GHG Protocol Corporate Value Chain (Scope 3) Standard (2011)
8. ISO 14064-1, Greenhouse gases—part 1 Specification with guidance at the organization level for quantification and reporting of greenhouse gas emissions and removals (2018)
9. Brian D, McGrayA H (2004) Comparison of ISO 14064 Part 1 and the GHG protocol corporate module
10. Hill N, Bonifazi E, Bramwell R, Karagiannim E (Ricardo Energy & Environment), Harris B (WRAP) for the Department for Business Energy & Industrial Strategy (BEIS) (2018) 2018 government ghg conversion factors for company reporting, methodology paper for emission

factors: final report, published by for the Department for Business Energy & Industrial Strategy (BEIS)

11. Scalbi S, Buttol P, Reale F, Masoni P (2016) Development of national databases of greenhouse gases emission factor. In: Proceedings of the 10th Italian LCA Network Conference "Life Cycle Thinking, sostenibilità ed economia circolare", Ravenna, 23–24 June 2016, ISBN: 978-88-8286-333-3 (2016), 447–455

12. World Resources Institute and World Business Council for Sustainable Development, GHG Protocol Product Life Cycle Standard (2011) Available at: http://www.ghgprotocol.org/com panies-and-organizations. Accessed 16 July 2019

13. Intergovernmental Panel on Climate Change (IPCC) (2006) IPCC Guidelines for National Greenhouse Gas Inventories. Available at: http://www.ipcc-nggip.iges.or.jp/public/2006gl/. Accessed 16 July 2019

14. EC (2013) European Commission, Commission Recommendation of 9 April 2013 on the use of common methods to measure and communicate the life cycle environmental performance of products and organisations

15. European Commission—Joint Research Centre—Institute for Environment and Sustainability: International Reference Life Cycle Data System (ILCD) Handbook—General guide for Life Cycle Assessment - Detailed guidance. First edition March 2010. EUR 24708 EN. Luxembourg. Publications Office of the European Union; 2010, ISBN 978-92-79-19092-6, ISSN 1018-5593 https://doi.org/10.2788/38479

16. Institute for Environmental Protection and Research (ISPRA), Italian Greenhouse Gas Inventory 1990—2015. National Inventory Report 2017. Available at: http://www.isprambiente.gov. it/it/pubblicazioni/rapporti. Accessed 16 July 2019

17. Institute for Environmental Protection and Research (ISPRA) (2015) Fattori di emissione atmosferica di CO2 e sviluppo delle fonti rinnovabili nel settore elettrico. Available at: http:// www.isprambiente.gov.it/it/pubblicazioni/rapporti. Accessed 16 July 2019

18. Institute for Environmental Protection and Research (ISPRA) (2016) Road transport database. Available at: http://www.sinanet.isprambiente.it/it/sia-ispra/fetransp/. Accessed 16 July 2019

19. Food and Agriculture Organization of the United Nations (FAO) (2015) Global database of GHG emissions related to feed crops. Available at: http://www.fao.org/partnerships/leap/dat abase/ghg-crops/en/. Accessed 16 July 2019

20. Fantin V, Scalbi S, Ottaviano G, Masoni P (2014) A method for improving reliability and relevance of LCA reviews: the case of life-cycle greenhouse gas emissions of tap and bottled water. Sci Total Environ 476–477:228–241

21. Patrizi N, Marchettini N, Niccolucci V, Pulselli FM (2018) Steps towards SDG 4: teaching sustainability through LCA of food. In: Proceedings of the 12th Italian LCA Network conference "Life Cycle Thinking in decision-making for sustainability: from public policies to private businesses", Messina, 11–12 June 2018, ISBN: 978-88-8286-372-2, 64–71

Carbon Footprint Estimation of an Indian Thermal Power Plant Towards Achieving Sustainability Through Adoption of Green Options and Sustainable Development Goals (SDGs)

Debrupa Chakraborty

Abstract Low carbon economy deserves special attention in connection to sustainability issues. Changeover to low carbon economy is possible with the reduction of carbon emission. This reduction mandate has nowadays become a 'global' goal keeping in view the catastrophe associated with the problem of climate change. This global goal can be achieved through innovations. Manifestation of low carbon goods leading to reduced emissions can be achieved with innovative products and production process. However, impact caused by industries on non-renewable natural resources and energy consumed in their production process and the resultant CO_2 emissions also need to be addressed and estimated. Technologies for generating electricity irrespective of its nature leads to generation of carbon dioxide (CO_2) and other greenhouse gas emissions. Carbon footprint (CF) is recognized as a method of evaluating carbon emissions in quantitative terms (expressed in tCO_2e) to respond to the increasing concern about climate change. However, it is necessary to make a life cycle inventory analysis by taking into account all the activities and processes that determine the size of carbon footprint. In this given backdrop, Carbon Footprint (CF) resulting from electricity generation in a coal-based power plant in India, for the year 2016–2017, has been studied based on primary data. Estimation of carbon footprint has been done using the Life Cycle Assessment (LCA) approach, been internationally accredited by ISO14044:2006 and ISO 14067:2018 and developed to understand the effects of individual products or process on environment. The cradle to grave approach of LCA has been followed so as to account for the CF resulting from the early stage of resource abstraction and its processing, production, delivery, disposal and reprocess or recycling. LCA approach method is apt for CF calculation or estimation at the micro-level. Analysis of result showed that that burning of coal occupies the lions share in the total carbon footprint of the power plant. Also based on the analysis, CF generated provided an useful means of guiding the power plant to distinguish or detect the 'hotbeds' of production process having an adverse bearing on the environment. The low carbon initiatives or green options adopted by the power

D. Chakraborty (✉)
Department of Commerce, Netaji Nagar College, Kolkata 700092, India
e-mail: chakraborty_debrupa@yahoo.com

© The Author(s), under exclusive license to Springer Nature Singapore Pte Ltd. 2021
S. S. Muthu (ed.), *LCA Based Carbon Footprint Assessment*, Environmental Footprints
and Eco-design of Products and Processes, https://doi.org/10.1007/978-981-33-4373-3_5

plant to reduce negative environmental impacts have also being discussed. Also, an effort has been made in this chapter to identify the Sustainable Development Goals (SDGs) impacted and achieved by the thermal power plant based on LCA results and adoption of green options. Finally, based on the results and low carbon options exercised by the unit, recommendations of possible green alternatives for achieving transition towards enhanced sustainability are also prescribed.

Keywords Carbon footprint (CF) · Life cycle assessment (LCA) · Indian thermal power plant · Green options · Sustainable development goals (SDGs) · Sustainability

1 Introduction

Low carbon economy deserves special attention in connection to sustainability issues. Changeover to low carbon economy is possible with the reduction of carbon emission. This reduction mandate has nowadays become a 'global' goal. This global goal can be achieved through innovations. With innovations, production of low carbon goods with reduced emissions can be manifested. This is again truer for developing nations like China, Brazil, India, etc. because they contribute significantly to the world economy. At the same time, their scope of mandated reduction of emission is much more than their developed counterparts.

World Resources Institute estimated India's greenhouse gas (GHG) emissions to be 6.5% of global total (value for the year 2014). India's, coal consumption increased by 4.8% or 27 million tonnes in 2017 and CO_2 emissions rose by 4.8% in 2018, principally driven by emissions from power plants based on coal [1]. The growth in the country's energy demand in 2018 was led by coal (for power generation) and oil (for transport). As a part of the Paris Agreement, India has put its best foot forward to achieve reduction in emission through Nationally Determined Contributions (NDCs) goals. India has set two set of targets—conditional and unconditional (by ratifying the Paris Agreement), which it aims to achieve through NDCs. The country has vouched to reduce its intensity of emission relating to GDP by 30–35% by 2030 (in comparison to the level in 2005) as a part of unconditional target. Along with this, it has also made a conditional target of generating 40% power from non-fossil-based sources by 2030 as a part of the country's ambitious NDCs. These targets have been put forward to the United Nations Framework Convention (UNFCCC) on Climate Change for a global climate accord in 2015 (forming a part of the Paris Agreement) (Paris Agreement, https://unfccc.int/process-and-meetings/the-paris-agreement/what-is-the-paris-agreement) [2].

Electricity generation technologies lead to the generation of carbon dioxide (CO_2) and other greenhouse gas emissions in respective of its nature. However, it is necessary to make a life cycle inventory analysis by taking into account all the activities and processes that determine the size of carbon footprint. Emissions can be both direct and indirect. The direct emissions relate to the operation of the power plant. On the contrary, the indirect emissions are related to the other non-operational phases of the life cycle. The largest carbon footprints have been estimated from fossil-fuelled technologies based on coal, oil, gas combustion. Carbon footprint (CF) is recognized as a method of evaluating carbon emissions in quantitative terms (expressed in tCO_2e) to respond to the increasing concern about climate change. One of the most important contributors towards GHG emissions is power plants generating electricity by fossil fuel like coal and oil. The smaller and older plants built before modern emission standards are the worst performers.

In the year 2018, it was observed that global atmospheric concentrations of carbon dioxide increased by 1.7%, which is 70% higher than the standard increase witnessed after 2010. According to IPCC, it has been found that 37% of the global emissions account to electricity production. It has been estimated that over the next 20 years, the electricity demand may increase by 43% [3]. An impressive growth from 1,713 MW to 344002.39 MW in the last few decades has been documented in the power generation sector in India. The major share of electricity generation can be attributed to coal-based thermal power plant, which generates a total of 197171.5 MW [1]. Hence, improvement in thermal efficiency of power plants may reduce CO_2 emissions in the country, which has become the need of the hour [1].

Joe Phelan, the India Director of World Business Council for Sustainable Development (WBCSD) believes that India has the potential to lead the inevitable transformation of the present business-as-usual scenario to net zero GHG emissions by 2050 and the achievement of Sustainable Development Goals (SDGs). But to exercise a limit on global warming to 1.5 °C above preindustrial levels require an amalgamation between transformative systemic changes and sustainable development. To achieve this 1.5° target and for maintaining peace and prosperity of the people and the planet now, and in future, all United Nations Member States in 2015 implemented the 2030 Agenda for Sustainable Development. The agenda constitute of 17 Sustainable Development Goals (SDGs) and appeal for action from both developed and developing countries to achieve these SDGs] (United Nations)] [4].

The 1.5 °C pathways specify strong synergies, particularly for the SDG 3 (Good health & wellbeing); SDG 7 (Affordable and clean energy), SDG 12 (responsible consumption and production) and 14 (Life below water) These SDGs are considered to be of very high confidence. For SDG 1 (No poverty), SDG 2 (Zero hunger), SDG 6 (Clean water and sanitation) and SDG 7 (Affordable and clean energy), SDG 13 (Climate Action), there is a danger of adverse effect from rigid mitigation actions compatible with 1.5 °C of warming. This demands unmatched liaison and initiative among businesses, investors and government [5]. Furthermore, the demand for energy is supposed to get doubled by 2040 due to increasing appliance ownership and cooling needs [6]. In this given scenario, Indian companies need to take up proactive role to focus on eco-efficiency and achieve SDGs, which is considered to be the first step towards 'Strong' Sustainable Development. In this given scenario growing number of companies are trying to adopt strategies compatible with SDGs and to support some or all of the SDGs. Companies are expected to show how far their existing product or service contributes towards supporting SDGs.

The remaining part of the paper is structured in four parts. Section 2 comprehends review of literation, Sect. 3 focuses on the goals and objectives of the present study, Sect. 4 explains the research methods and data sources, Sect. 5 provides the results obtained from thermal power plant. Section 6 discusses the green options adopted by the case study unit and Sustainable Development Goals (SDGs) achieved as a result of adopting the green options. Finally, Sect. 7 concludes the chapter with some recommendations that the power plant can adopt as its future pathway for achieving sustainability.

2   Review of Literature

CF is widely accepted as a method of quantifying greenhouse emissions (GHG) of a product or process during its entire life cycle. In this section, review of literature is conducted on carbon footprint (CF) estimation of industries (using non-renewable and renewable sources of energy) in India and abroad based on various methodologies of calculating CF. In 2002 in China, with the aim to make provision of 1kWh of electricity for all consumers, Life cycle inventory (LCI) was quantified for identifying areas of improvement in the industry [7]; Life-Cycle Assessment (LCA) methodology was utilized in the industrial level for estimating GHG emissions in coal-based power stations [8]. Later, the emissions from the coal-fired power stations of China were analyzed using Life cycle accounting model [9]. Studies also exist relating to estimation of greenhouse gas (GHG) emission of wind electricity in Europe using LCA approach [10]. A Hungarian biogas power plant estimated carbon footprint and the associated stability of energy using complete life cycle assessment of biogas production. Results showed that 50% of the valuable energy is wasted to measure the needs to be taken to utilize such valuable energy [11]. Study based on Indian power plants revealed measuring GHG emissions based on methods other than LCA has been made. Emission from purely coal-based generation plants without using any kind of non-renewables was studied [12]. Also application of renewable like sawdust along with coal for generation of electricity was studied for an Indian thermal plant to reveal the potential option for the energy development programme [13]. Impacts of using biomass feed-stocks as fuel in gasifiers of Indian thermal power plants have been done in another work and the bio-energy generated has been estimated [14]. Further literature review reveals a variety of LCA studies across the globe that includes—16 different studies related to LCA of the electric grids and their components were reviewed to compute the impacts on environment due to grid losses and life-ending condition of grid components [15]. CF estimate based on life cycle assessment and its related environmental performance assessment, was undertaken for an electricity grid in Greece using different power generation technologies, [16]; calculation of CF and its environmental significance using different combinations electricity generation by 2050 was studied for Sri Lanka [17]. However, since the inception of SDGs in 2015 and also given its gaining importance for businesses, developing a link between LCA and SDGs has become inevitable. Thus review of studies in this line also has been made. Sustainable development goals were used as a guideline for assessing sustainability with the help of Life Cycle Assessment approach for a electrolytic hydrogen manufacturing [18]. A more updated recent draft prepared by the United Nations Life Cycle Initiative has established a link between the United Nation Sustainable Development Goals and the pathways in which life cycle create an impact. This draft was prepared based on a project to create robust links between the SDGs and LCA and to develop a methodology for measuring and reporting on companies' contributions to the SDGs [19].

Various methods of calculating CF of product already exist as revealed in the review of literature, though variations exist in manner or form of their implementation or operation. Amongst the several models for measuring Product Carbon footprint (PCF), Life-Cycle Assessment (LCA) model is one of the most standardized and widely used ones. The outline or framework of a variety of LCA models has been prescribed by a number of organizations and bodies of international repute namely Publicly Available Specifications (PAS) 2050 by British Standard Institute in 2008 and revised in 2011; Greenhouse Gas (GHG) Protocol by World Resource Institute(WRI) and World Business Council for Sustainable Development (WBCSD) in 2011; ISO 14067 Carbon Footprint of product by International Organization for Standardization in 2012 and revised version in 2018 [20]. PAS 2050 helps towards quantification of CF at a product level, GHG Protocol requires quantification and public reporting of GHG inventories of products. ISO 14067 based on LCA Standards ISO 14040 and 14044 emphasizes on quantifying and reporting CF of product addressing the issue of impact on climate change [21].

Review of literature reveals that though greenhouse gas (GHG) emissions have been estimated for a number of industries based on various methodologies including LCA approach, but studies showing impact of carbon—di—oxide (CO_2) on SDGs at each stage of the LCA is still at a rudimentary stage and hard to find specially for Indian industries. This issue needs to be addressed given the rise in emissions propelled by coal-fired power plants in a developing economy like India and importance of achieving SDGs by resource-intensive and GHG emitting industries.

3 Goals and Objectives

The goal of this study is the life cycle assessment of a non-renewable resource (coal and oil)-based new power plant to quantify CO_2 emission resulting at the different stages of its life cycle. The emission is being assessed in terms of carbon dioxide and expressed as ton carbon equivalent (tCO_2e). The results of the study aim to identify the 'hotbeds' so that steps can be taken to achieve reduction in carbon—di—oxide (CO_2) emissions and also to make an assessment of the impact of such emissions on SDGs at each stage of the LCA. The low carbon initiatives or green options adopted by the power plant to reduce negative environmental impacts and attainment of SDGs by the unit have also being discussed. Finally, based on the results and low carbon options exercised by the unit, recommendations of possible green alternatives are also prescribed.

4   Research Methods and Data Sources

Research Methods

Life-Cycle Assessment (LCA) Approach has been adopted as this methodology helps to quantify carbon footprint of product or processes with considerable accuracy and allows organizations to operate with much adaptability. ISO 14067:2018, an International Standard based on globally acceptable principles, requirements and guidelines for the quantification and reporting of the carbon footprint of a product (CFP) along its life cycle has been used in this study. This newly upgraded standard helps organizations of all kind to use this method to calculate the carbon footprint of their products and also to understand the approaches of CF reductions in a better way. This approach is a key way of contributing to the achievement of international climate action goals (ISO 14067:2018 https://www.iso.org/standard/71206.html) [20]. CF data are used for decision-making purpose although it neglects the other important environmental impacts, for example, acid rain, smog, cancer effects and land use. Consideration and evaluation of these environmental impacts will ensure a more complete LCA.

LCA utilizes some of the common factors in this study that are enumerated below:

- Functional unit
- Study Boundary
- Life cycle inventory (LCI) data sources
- Environmental impact categories (Midpoint and Endpoint)
- Interpretation, reporting, peer review

The complete methodology constitutes of LCA evaluation and the process can be subdivided into the following four phases [22, 23]:

1. Goal and Scope definition—Attempts to address two questions; purpose of study (why) and the targeted group (for whom).
2. Inventory Assessment or Life Cycle Inventory (LCI)—In this phase, demarcation of boundary, identification of functional unit and collection of data are undertaken. Collection, recording and data measurement in connection to the input items such as energy, water and raw materials while the output based on the products formed, emissions to air, water and soil is performed. The data stated are in reference to the functional unit and are expressed as a common unit.
3. Life Cycle Impact Assessment (LCIA)—Environmental impact categories or interventions (Midpoint and Endpoint) are evaluated at this phase. Environmental interventions include—raw material extraction, emissions (air, soil, water) and impact on land and biodiversity.

Mid point interventions include impact categories relating to climate change, land use, water use, resource depletion, biodiversity, toxic effects on human life etc.

End points interventions include damage categories relating to casing negative effect on human health, resource depletion and quality of ecosystem.

4. Interpretation—This is the last phase of LCA study and interpretation at this stage is to be to detect the new stretches and collect related new data that have prospects for improvement. Based on these interpretations, commendations are made at this phase.

In this connection choosing boundaries is precisely significant for LCA activities. This is true for any process or product. The different boundaries can be of the following four categories [23]:

- Starting from supply items at mining site (cradle) to recycling of products after being sold off to end consumers, 100% consumption of wastes (cradle).
- Commencing from extraction of resource (cradle) to final discarding of generated wastes (grave).
- From raw materials abstraction (cradle) to transportation of final products to factory gate (gate) before sending final products to the customers. This process excludes the use and discarding of wastes by the unit.
- The pathways relating to processing within the whole production chain (gate) inside the factory premises (gate).

LCA assessment method has being adopted as this method can help to move towards Sustainable Development Goals (SDGs). This has gained prominence all more since the United Nations Environment Progrmme (UNEP) pioneered a Life Cycle Initiative project establishing a link between the UN Sustainable Development Goals (SDGs) and life cycle assessment (LCA) in 2018. This is because since the inception of SDGs in 2015, companies initiated processes showing their willingness to achieve these goals. However, the aims and indicators of SDG relevant for businesses are still at a very nascent stage, so it is a challenging task to quantify and supervise the results effectively on the part of the companies. However, an effort has been made to identify the SDGs impacted and achieved by the thermal power plant based on LCA results and adoption of green options.

Functional Unit

The functional unit is a coal- and oil-based Indian thermal power generating unit. Considered as one of the most renowned coal-based power plants in India, the total generation capacity of the plant is 1050 Megawatt (MW) whereas the total generating capacity of the unit is 7300 Million Unit (MU) of electricity. During the year under consideration, i.e., 2016–17, the total electricity generated from the power plant was measured to be 7050 Million Unit (MU) (1 MU = 10,00,000 kWh of generated electricity).

Study Boundary

Study boundary encompasses—organizational boundary and operational boundary. Attempt has been made to keep the organizational boundary as broad as possible keeping focus on the GHG emissions (only CO_2 emissions) emitted by the facilities during power generation. The operational boundary embraces extraction of resource (cradle) to final discarding of generated wastes (grave).

The total process necessary to generate electricity constitute—transportation of coal (hard Grade D/E), fuel (Light Diesel Oil or Heavy Fuel Oil) and other chemicals that are present in the factory premises; the production process and other service-related activities carried out at the administrative office, employees business tours; and at the final stage transportation and disposal of wastes, for measuring the total carbon emission of the power plant unit. Emission of carbon at each stage of the total process was calculated. The emission factors are mainly calculated based on the secondary sources.

For estimating emission from transportation of coal, fuel, generated waste and employees travel by road, India Specific Road Transport Emission Factors, Version 1.0. (under India GHG Program prescribed by WRI India, TERI and CII in 2015) [24] was used. On the other hand, India Specific Railway Transport Emission Factors, Version 1.0. (under India GHG Program prescribed by WRI India, TERI and CII in 2015) [25] has been used for estimating emission from transportation of coal, fuel oil and chemicals by rail.

Also, emission factor approved by Guidance for Voluntary Corporate Greenhouse Gas Reporting—2015 [26] has been followed for assessing emission from coal and fuel consumed in the boiler for generating electricity. CF has been calculated at each stage and summation of all the stages provided the total CF of the unit. The activity data (kg/litres/kWh/tkm) at each stage of the LCA are multiplied by the emission factor (kg CO_2e per kg/litres/kWh/tkm). However, impact of manufacture of equipment and chemicals used in the power generation process has been excluded.

For all three stages, emission from water used during the generation process was taken into account. As a part of eco-friendly drive rainwater from the two nearby dams are used by the plant to meet its total water consumption. Also a considerable portion of the water used is treated, examined and finally discharged to the nearby river.

Data Sources

The study was carried out at the power plant of the said company. A questionnaire was prepared beforehand on the basis of which primary data were obtained verbally from the plant engineer, boiler house operators and the store manager after repeated visits to the plant. The employees working on the shop floor were also consulted as they are involved in the day to day activities of the plant, and that proved to be useful.

CF of the plant has been estimated based on the primary data and using emission factor from secondary sources. CO_2 emissions from both the direct and indirect sources of energy were calculated for 2016–2017. Direct emission resulting from the stack of power plant, operation of power plant was calculated. Indirect emission from both upstream and downstream was taken into consideration. Upstream provisions include fuel, resources required for transportation of raw materials, employees business travel/tours, energy requirement in administrative office, whereas downstream provisions include emission resulting from transportation of wastes and by-products to the related sites, agents and companies.

Limitations of Study

However, a number of problems were faced while carrying out the study. The supply of coal at the power stations was irregular at certain point of times, which disrupted the production process and made the data collection problematic; data relating to certain aspects were confidential in nature and access to such data could not be made.

5 Results

The detailed estimation CF based on LCA approach has been shown in Table 1. The basis of estimation of CF at each of the three stage of life cycle of the plant is explained and finally the total CF and the total emissions from each of the direct and indirect sources have been explained in detail.

A. *Carbon Footprint from transportation of coal, fuel, chemicals (Indirect Emissions)*

The requirement of raw materials for the plant is met by a number of suppliers. The supply of the key raw material for production, coal is obtained from Coal India Ltd. The company provides around 65% of the total coal requisition of the plant. A number of coal agencies namely Mahanadi Coal Fields Limited (MCL), Bharat Coking Coal Limited (BCCL), and Eastern Coal Fields Limited (ECL) are the suppliers of 28% of the required coal and the rest 7% of the coal is gathered from e-auction. For calculating emission from combustion of coal, it is assumed that non-coking coal (Grade D and E) was utilized. In 2016–17, the quantity of coal and light diesel oil that were consumed has been found to be respectively 45,46,282 Metric Tonnes (MT) and 5,350 kilo litres (KL). Supply of coal was brought mainly by rail route from varied distant places (stretching from 20 km to 250 km). So the annual coal requirement of 45,46,282 MT of coal is transported from an average distance of approximately 75–150 km.

This is based on an approximate estimation of 12,500 MT of coal per day for 365 days in 2016–2017 to the power plant from several destinations by rail route. On an average 12,500 MT of coal was transported by five trains per day. Emission factor of 0.0095 kgCO$_2$e/tkm [India Specific Rail Transport Emission Factors, Version 1.0. under India GHG Program prescribed by WRI, TERI and CII, India in 2015) has been taken for calculation of emission from coal combustion.

Table 1 Total CO_2 emissions (carbon footprint) of Indian Power Plant (2016–2017)

Transportation/Production components	Quantity of resources/Services used	Unit of resources/Services used/Consumed	Secondary emission factor	CO2 emission (tCO$_2$e)
Transportation of Coal by railways	2,851,562,500	tkm	0.0095 kgCO$_2$e/tkm	27090
Transportation of light fuel oil by railways	1200	Km	0.7353 kg CO$_2$e/km	1
Transportation of chemicals used in production process by road, 100% loaded freight trucks/lorries	1,56,750	tkm	0.143 kg CO$_2$e/t km	22
Sub-total (A)				27,113
Bituminous coal used in boiler for generation	45,46,282	MT	2.65 kgCO$_2$e/kg	120,47,647
Fuel used for generation	5,350,000	Litres	2.95 kgCO$_2$e/litre	15,783
Sub-total (B 1)				120,63,430
Business Travel/Tours by the Employees: Two Wheelers (Motorbikes and Scooters)	33,00,000	Passenger kms	0.0387 kgCO$_2$e/pax-km	128
Four wheeler (cars)	13,20,000	Passenger kms	0.111 kgCO$_2$e/pax-km	147
Sub-total (B2)				275
Sub-total (B1 + B2)				120,63,705
Transportation of wastes by road, in 100% loaded freight tractors/lorries)				
Wastes (Rejected materials) generated in mills	14,62,000	MT	1.13 kg CO$_2$e/kg	1652
Wastes Disposed: (Dry Ash, Wet Ash and wastes generated in mills are recycled by end users)		–	–	NIL
Sub-total (C)				1652
Total (A + B + C)				120,92,470

The total annual supply of light diesel oil (5,350 KL) was met by oil refineries located at a distance of 550–600 kms from the power plant. The oil is transported by rail route twice a year. So for fuel oil transportation by rail, a distance of approximately 1200 km and emission factor as per India Specific Rail Transport Emission

Factors under GHG Program, 2015, Version 1.0. is considered for calculation of emission.

Chemicals required annually (around 1650 MT) for production purpose were transported by road from suppliers based in locations ranging between 25 and 250 km. For calculation of emission from transportation of chemicals emission factor of freight vehicle as per India Specific Road Transport Emission Factors, Version 1.0. (under India GHG Program prescribed by WRI India, TERI and CII in 2015) has been used.

Overall carbon footprint from transportation of coal, fuel and chemicals in 2016–17 amounted to 27,113 tCO_2e.

B. *Carbon Footprint of production process, energy used in administrative offices and business travel/tours of employees (Direct Emission)*

The total annual electricity generated by the coal-based plant (constituting of five generation units each having a capacity of 210 MW) was 1050 MW. The coal used in the production of electricity is non-coking bituminous coal having 35% ash content. A quantity of 45,46,282 MT (of grade D and E) was fed into the boiler for achieving higher efficiency. This was supported by use of 5,350 kilo litres (KL) of Light Diesel Oil. Estimation of CF produced by combustion of coal and oil has been made based on emission factors prescribed by Guidance for Voluntary Corporate Greenhouse Gas Reporting, 2015 by Ministry of Environment exclusively for Indian industries using Data and Methods from the 2013 calendar (INFO 734) was used. The auxiliary operations of the plant consumed 9.56% of the total annual generation of 7050 million kWh in 2016–2017. This auxiliary energy requisite is met out the power generated by the plant. So CF from production of electricity amounted to 120,63,430 tCO_2e.

Another form of direct emission resulted from the energy used in the administrative offices of the plant and from the business travels of the officers and other employees of the plant and its office. The energy requirement of the office is met out of generated electricity, so emission from this source is not measured independently.

So far as emission from business travel of the officials are considered a number of information was obtained from the travel register of the plant and the distance travelled was calculated on the basis of the gathered information and assuming that the two wheelers used have an engine capacity of 1500 cc and more and four wheelers are of less than 1000 cc capacity (small cars). CF of business travel was made based on India Specific Road Transport Emission Factors, Version 1.0. under India GHG Program prescribed by WRI India, TERI and CII in 2015. Travel register information revealed the details of the number of two- and four-wheeler vehicles and their corresponding travel days and distance travelled 2016–2017:

(i)   Two wheelers (around 500 diesel-operated vehicles) run on an average between 10 and 20 km per day for 300–340 days.

(ii) Four wheelers (around 35–40 diesel-operated small cars and small buses) travel an average distance of 70–100 km per day for 300–340 days.

(iii) 500 bicycles are on run for 330–365 days travelling a distance of 3–5 km per day.

Total emission from travel in two wheelers (scooters and motorbikes, mainly owned by the employees) amounted to 128 tCO$_2$e and that from commuting in four wheelers (employee owned cars owned and or hired vehicles like buses) was found to be 147 tCO$_2$e.. Emission from commuting by bicycle is zero. So the total CF from business travel by officers and other employees totalled to 275 tCO$_2$e. Annual CF from production process and business travel amounted to 120,63,705 tCO$_2$e in 2016–17. The footprint at this stage of the life cycle of the plant forms the largest part of the annual CF and is the 'hotbed' having a significant impact of the environment.

C. *Carbon Footprint from transportation and disposal/dumping of wastes (Indirect Emission)*

Waste generated constitutes of fly ash (dry ash formed from burning of coal), wet ash, and rejected materials like stones, heavy minerals etc. Both dry and wet ash handling system is used by the plant for transportation of the generated wastes. Generally, dry ash handling system is used for transportation of fly ash, which is transported in closed containers by lorries or dampers for the purpose of environmental protection. Normally such wastes are transported around a distance of 35–45 km per day and are sold off to the municipality, traders and to cement manufacturing industries. About 18,06,973 MT fly ash is sold off at Rs. 90–95 per ton (generating a revenue of Rs.16 to 17 crores in 2016–2017) to the local municipal authority, vendors and brick and cement manufacturing industries who collected and transported the fly ash in their own vehicles, bearing all the associated cost.

About 7,85,820 MT (2150 MT per day) of wet ash is collected by traders and local municipal authority free of cost. The vendor and municipal authority transports the wet ash over a distance of normally 6–10 kms per day by lorries and get them recycled by evacuating them for reclamation of low lying land and construction of roads in neighbouring region. Since these wastes are collected by the traders from the premises of the plant and not transported by the plant itself, emission from transportation of dry and wet ash waste is considered to be zero.

Around 80 MT–100 MT of rejected materials gets normally generated per day from production process. These rejects are transferred to the end users for recycling purposes. Generally around 15–20 tractors/lorries are loaded with these rejects and are transported by the plant to the end users in the neighbouring areas (a distance of 2–5 kms on an average daily). A total of 14,62,000 MT rejected materials were generated and for calculating the emission from transportation of these rejected materials the default emission factor (1.13 kg CO$_2$e/kg) relevant for India was considered or used (India Specific Road Transport Emission Factors, Version 1.0. under India GHG Program prescribed by WRI India, TERI and CII, in 2015). Calculation revealed that the total CO$_2$e emission from transportation of wastes was to the tune of 1,652 and that from disposal of wastes was zero (all the generated wastes been recycled) for the year 2016–17.

D. Total/Annual Carbon Footprint

Total CF based on the summing up of emissions at each stage of the life cycle of the power plant following LCA approach for the year 2016–2017 was 120,92,470 tCO_2e. This emission resulted from annual generation of 7050 million kWh of electricity. Direct emission (120,63,705 tCO2e) resulting from burning of coal for production operations and business travel by employees and officers constitutes the lions share (around 99%) in the total CF and indirect emissions (28,762 tCO2e) constituting of upstream and downstream emissions made a minor contribution of one per cent therein.

Upstream indirect emissions (27,112 tCO2e) encompass emission resulting from transportation of raw materials like coal, fuel, chemicals, business journeys of officers/employees, energy used for running the offices for administrative purposes.

Downstream indirect emissions (1,650 tCO2e) constitute of transportation of wastes to various agencies and industries for recycling and transporting by-products to various end users and companies.

Based on these results, the green options adopted and the SDGs impacted and achieved by the plant have been discussed in the following section for addressing GHG reductions and associated climate change issues.

6 Green Options Adopted by the Unit and SDGs Achieved

From the CF results obtained, it is understandable that GHG emission reductions across life cycle of the production process need to be addressed through mitigation pathways so as to reduce negative impacts of climate change. Deferring climate mitigation and adaptation responses would lead to an adverse environmental impact and relegate the vision of sustainable development. In this given backdrop, based on the results of annual CF, environmental impact made by the power plant at each stage of LCA and the extent to which LCA impact categories contribute towards the SDGs involved has been projected in Table 2.

Also green options adopted by the plant through energy conservation to contribute towards lowering of CF and through ash handling to reduce negative environmental effects are discussed below. Green options have been adopted through energy conservation to contribute towards lowering of CF and through ash handling to reduce negative environmental effects. The unit invested Rs.13.5 crores towards green initiatives in 2016–17.

Energy Conservation

- Auxiliary Power Consumption (APC) was reduced to 9.56% by tap changing operation of transformers.
- Old Capacitor Voltage Transformer (CVT) of low accuracy was replaced by 0.2 classes CVT and replacement of cartridges resulted in the reduction of current consumption by 10 Ampere (AMP) and power consumption by 100 KWh.

Table 2 Impact on sustainable development goals (SDGs) at each stage of lca of an indian power plant (2016–2017)

	LCA stage	LCA environmental impact categories	SDGs impacted	Nature of SDG
A	Carbon Footprint from transportation of coal, fuel, chemical by railways	1. Climate (emission associated with transportation of raw materials)	SDG-13	Climate action
B.	Carbon Footprint of production process, business travel/tours of officers/employees and energy consumed for operations at administrative office	1. Climate 2. Biodiversity (is effected because of combustion of coal and emissions to air)	SDG-3	Good health and well-being
			SDG-7	Affordable and clean energy
			SDG-11	Sustainable cities and communities
			SDG-12	Responsible Production and consumption
			SDG-13	Climate action
C	Carbon Footprint from transportation of Waste Materials	1. Climate (emission associated with transportation of waste)	SDG-13	Climate action
	Carbon Footprint produced during disposal of Wastes	2. No impact as all three types of generated wastes are recycled by end users	–	–

- Power to the local township was provided from the installed rooftop Photovoltaic (PV) panels.
- Online monitoring of flue gas, monitoring of water quality of ash pond was found functioning successfully.
- Coal Dust Injection (CDI) was installed to convert coal into fine dust to improve the environment in coal mines & coal yard area.
- Energy Efficient luminaries were introduced in place of Old Energy-intensive luminaries.
- Photosensors have been installed in different rooms and conference hall for auto switch on/ off of lights.

- Under PAT[1] Cycle-II, the unit achieved Specific Energy Consumption (SEC) and has earned 21,333 Energy Saving Certificates (ESCert) during the year 2016–2017 (one ESCert is issued for saving one ton of oil equivalent saved by the unit in the process). The price of each ESCert is equal to 10,498 INR.

Ash Handling/Utilization and Environmental Management: the plant operation results in two types of ash; Dry and Wet

- The company was found continuously making efforts to use fly or dry ash. The fly ash generated (18,06,973 MT in 2016–17) was sold off to various organizations and industrial houses for recycling purposes through MOU. These organizations recycle the dry ash into environment friendly ash slabs, ash tiles and for construction or erection of roads. This whole process of recycling contributes towards generation of revenue as well. The Company was observed taking initiatives to start its own brick making unit from 2017 to 2018 and had already purchased machine for brick making.
- The wet ash generated during the production process are dumped into ash pond created for the purpose such disposal so as to reduce environmental pollution. However, it was necessary to initiate dialogues with Commerce & Industries Department, Government of West Bengal for introducing incentives schemes for setting up of new brick manufacturing and ash-based industries utilizing the said ash.
- Silencers for start-up vents and ejectors were installed to eliminate sound pollution.
- Installation of Scrubbers was found resulting in reduction of Sulphur Oxides (Sox), Nitrogen Oxides (NOx) emission rate from 100 to 30 over a period of last 5 years.

The adoption of energy conservation measures, ash handling and environmental management initiatives have helped the power plant to achieve the environmental SDGs namely SDG 3—Good Health and Well Being, SDG 7—Affordable and Clean Energy, SDG 9—Industry, Innovation and Infrastructure, SDG 12—Responsible Production & Consumption, SDG 13—Climate Action. The adoption of green

---

[1]National Mission for Enhanced Energy Efficiency (NMEEE) introduced the scheme of Perform Achieve and Trade (PAT) for attaining reduction in energy consumed by industries in which demand for energy is very high. However this reduction has to be achieved by such industries through certain market based mechanism that would improve the cost efficiency by trading of certificates received by industries for saving excess energy.
The first cycle of Perform Achieve and Trade (2012–13 to 2014–15) was framed with the intention of reducing specific energy consumption (SEC) in energy intensive sectors. There are 478 Designated Consumers (DCs) from eight sectors namely, Aluminum, Cement, Chlor Alkali, Fertilizer, Iron & Steel, Paper & Pulp, Thermal Power Plant and Textile. However PAT Cycle –II was formed for a duration of three years commencing from 2016–17 to 2018–19. In this second cycle of PAT, three new industrial sectors namely Railways, Refineries and Electricity Distribution Companies (DISCOMs) were incorporated. This inclusion was made for the purpose of accomplishing a reduction in total energy consumption of 8.869 Million Tonnes of Oil Equivalent (MTOE) (http://www.beeindia.gov.in) [27].

measures or options has led towards reducing the degree of impact on all the SDGs (as shown in Table 2) except that on SDG 11 (Sustainable cities and Communities). However, the extent or degree to which achievement of SDGs has been attained need to be measured and studied in future through more detailed research analysis.

Energy conservation measures will not alone be sufficient to tackle the problem of climate change produced by emissions from burning of coal. Coal-based generation should be supported by generation from alternative clean sources of energy. Then again, for production of power from alternative renewable sources of certain category will create a load on the limited resource like land. For example, the implementation of mitigation measures such as using land for growing bioenergy crops or taking measures like afforestation for creating carbon sinks for absorbing emissions will have a rebound effect. So land use and change in land use including land degradation and desertification (through measuring and monitoring) should be accounted for. This can be done through improved use of new knowledge and latest interaction technologies (cellphone-centred applications, cloud-computing etc.) [28].

7 Conclusions and Recommendations

The study revealed the process of estimating carbon footprint during the full life cycle of coal-based power generating unit in India. The total CF of the thermal power plant for the year 2016–17 was estimated at 120,92,470 tCO_2e. This emission occurred to generate electricity to the tune of 7050 million kWh. The CF calculation has helped towards identification of the hotspots/hotbeds of carbon emission thereby helping the unit to distinguish the areas of operation where green initiatives are to be implemented and also in identifying the Sustainable Development Goals (SDGs) impacted and achieved. Being a coal-based plant, direct emission comprised about 99% of total carbon footprint and indirect emission only as minor around 1%. The unit has undertaken a number of carbon reduction and other green options or initiatives as mentioned above along with measures like afforestation/tree plantations in the nearby locality.

However, it is further recommended that

- Milling system to be upgraded for grinding coal to desired sizes to make possible cent percent burning of coal and lessening of dry ash. Preference to be given to dry ash collection and disposal (compared to wet ash) because of dearth of land available for fly ash disposal.
- Renovation and modernization of the plant will help to lengthen the effective life span of the plant by at least two more decades.
- Introduction of renewable energy like solar, wind, biogas can help towards reduction of carbon footprint (direct emission) and to achieve SDGs in a more concrete way.
- Solar PV units may be introduced for Electricity generation to be utilized for lighting purpose.

- Carbon char may be introduced in place of some amount of coal to reduce the carbon footprint.
- Char may be utilized in place of oil support to keep flame stability in times of load fluctuations.
- Biofuel may also be utilized in place of oil support.
- Some renewable-based power can be generated or purchased for supporting auxiliary operations.
- Initiatives to be taken to make increase awareness amongst employees about environmental obligations, communicating the commitment towards SDGs (change in product development and reporting), scaling up measures for meeting the new challenge of using LCA-based metrics for SDGs.

References

1. Central Electricity Authority (2018) Ministry of Power Central Electricity Authority. Government of India, New Delhi Power Sector
2. Paris Agreement https://unfccc.int/process-and-meetings/the-paris-agreement/what-is-the-paris-agreement
3. World Nuclear Association (WNA) Report (2018) Comparison of lifecycle greenhouse gas emissions of various electricity generation sources
4. United Nations. (n.d.). (2020) Helping governments and stakeholders make the SDGs a reality. In: Sustainable development goals knowledge platform: stakeholders make the SDGs a reality
5. De Coninck HAC (2018) Strengthening and implementing the global response. In Masson Delmotte VP-O-O (ed) Global warming of 1.5 °C. An IPCC Special Report on the Impacts of Global Warming of 1.5 °C above Pre-industrial Levels and Related Global Greenhouse Gas Emission Pathways, in the Context of Strengthening the Global Response to the Threat of Climate Change. IPCC, 313–444
6. CDP (2020) Climate and business partnership of the future CDP India Annual Report 2019. CDP
7. Di X, Nie Z, Yuan B et al (2007) Life cycle inventory for electricity generation in China. Int J Life Cycle Assess 12:217. https://doi.org/10.1065/lca2007.05.331
8. Whitakar M, Heath GA, Donoughe P, Vorum M (2012) Life cycle greenhouse gas emissions of coal fired electricity generation: systematic review and harmonization. J Indus Ecol 16
9. Wang N, Ren Y, Zhu T, Meng F, Wen Z, Liu G (2018). Life cycle carbon emission modelling of coal-fired power: Chinese case. Energy 162:841–852. https://doi.org/10.1016/j.energy.2018.08.054
10. Pandey P, Blanc I, Le Boulch D, Xiusheng ZA (2012) Simplified life cycle approach for assessing greenhouse gas emissions of wind electricity. J Ind Ecol 16:S28–38
11. Szabó G et.al (2014) The carbon footprint of a biogas power plant. Environ Eng Manage J 13(11):2867–2874
12. Chakraborty N, Mukherjee I, Santra AK, Chowdhury S, Chakraborty S, Bhattacharya S, Mitra AP, Sharma C (2008) Measurement of CO2, CO, SO2, and NO emissions from coal-based thermal power plants in India. Atmos Environ 42:1073–1082
13. Sarkar S, Sahu SG, Mukherjee A, Kumar M, Adak AK, Chakraborty N, Biswas S (2014) Co-combustion studies for potential application of sawdust or its low temperature char as co-fuel with coal. Appl Therm Eng 63(2):616–623
14. Roy PC, Dutta A, Chakraborty N (2013) An assessment of different biomass feedstocks in a downdraft gassifier for engine application. Fuel 106:864–868

15. Gargiulo A, Girardi P, Temporelli A (2017) LCA of electricity networks: a review. Int J Life Cycle Assess 22(10). https://doi.org/10.1007/s11367-017-1279-x
16. Orfanos N, Dimitris M, Angeliki S, Dedoussis V (2019) Life-cycle environmental performance assessment of electricity generation and transmission systems in Greece. Renew Energy 139:1447–1462. https://doi.org/10.1016/j.renene.2019.03.009
17. Danthurebandara M, Rajapaksha L (2018) Environmental consequences of different electricity generation mixes in Sri Lanka by 2050. J Cleaner Prod 210. 10.1016/j.jclepro.2018.10.343
18. Wulf C, Werker J, Zapp P, Schreiber A, Schlör H, Kuckshinrichs W (2018) Sustainable development goals as a guideline for indicator selection in life cycle sustainability assessment. In: 25th CIRP life cycle engineering (LCE) conference. Elsevier, Copenhagen, Denmark, 59–65
19. Weidema B, Goedkoop M, Meijer E, Harmens R (2020) LCA-based assessment of the sustainable development goals: development update and preliminary findings of the Project "Linking the UN Sustainable Development Goals to life cycle impact pathway frameworks". PRé Sustainability & 2.0 LCA consultants
20. ISO 14067:2018 Greenhouse gases—carbon footprint of products—requirements and guidelines for quantification (2018) https://www.iso.org/standard/71206.html
21. Chakraborty D (2015) Product carbon footprint estimation of a tonne of paper: case study of a paper production unit in West Bengal, India. In: The carbon footprint handbook. CRC Press, Taylor & Francis Group, pp 487–502
22. Hague N (2016) Carbon footprint calculation and principles of life cycle evaluation for a bioenergy plant. In: 7th LCA Conference, Melbourne
23. Klemes JJ (2015) Overview of environmental footprints. Assessing and measuring environmental impact and sustainability, Elsevier, 131–193
24. India GHG Program. (2015) India specific road transport emission factors, Version 1.0. WRI India, TERI and CII
25. India GHG Program (2015) India specific railway transport emission factors, Version 1.0. WRI India, TERI and CII
26. Guidance for Voluntary Corporate Greenhouse Gas Reporting (2015) Using data and methods from the 2013 calendar by Ministry of Environment, INFO 734
27. Bureau of Energy Efficiency, Government of India, Ministry of Power http://www.beeindia.gov.in
28. IPCC (2019), Arneth et al (eds) Summary for policymakers. Climate change and land: an IPCC special report on climate change, desertification, land degradation, sustainable land management, food security, and greenhouse gas fluxes in terrestrial ecosystems

Printed in the United States
by Baker & Taylor Publisher Services